SpringerBriefs in Applied Sciences and Technology

PoliMI SpringerBriefs

Alessandro Di Pretoro · Flavio Manenti

Non-conventional Unit Operations

Solving Practical Issues

POLITECNICO
MILANO 1863

Alessandro Di Pretoro
CMIC
Politecnico di Milano
Milan, Italy

Flavio Manenti
CMIC
Politecnico di Milano
Milan, Italy

ISSN 2191-530X   ISSN 2191-5318   (electronic)
SpringerBriefs in Applied Sciences and Technology
ISSN 2282-2577   ISSN 2282-2585   (electronic)
PoliMI SpringerBriefs
ISBN 978-3-030-34571-6   ISBN 978-3-030-34572-3   (eBook)
https://doi.org/10.1007/978-3-030-34572-3

This Springer imprint is published by the registered company Springer Nature Switzerland AG
The registered company address is: Gewerbestrasse 11, 6330 Cham, Switzerland

Preface

This volume is the culmination of the authors' collaboration aimed to provide a reference book for Chemical Plants II course, addressed to the M.Sc. Chemical Engineering students at the Department of Chemistry, Materials and Chemical Engineering "Giulio Natta".

It was conceived as a quick and complete reference compendium of several non-conventional unit operations. This topic is discussed in specific optional courses but it is seldom explained during traditional lectures, although a chemical engineer often needs to deal with non-conventional unit operations during his career. Actual scientific literature includes excellent volumes which provide in-depth analysis of every single unit operations, but they require an extensive study to find the pursued information, an effort which could be useful only for highly specialized and sectoral applications. The first reason to write this book was indeed to meet the chemical engineering students' need for effectively achieving a fundamental knowledge of a wide range of non-conventional processes, without preventing further studies if required.

The text provides both methodologies and numerical applications for non-conventional unit operations design. Based on a proposed case study, process model equations are extrapolated and followed by the numerical solution in order to immediately test the acquired skills on an application example. At the end of certain chapter, a C++ code implementation is provided in order to solve the ODE or nonlinear equations system by mean of the ad hoc BzzMath library.[1] The secondary objective of this book is therefore to create a connection between the modeling problem and its computational solution, regardless of the coding language, that is the basic concept of Computer Aided Process Engineering.

The first part deals with design, comparison, and optimization of heating and cooling operations, different from simple heat exchange, which are widely used both in industrial and daily life settings.

---

[1]Downloadable at: https://super.chem.polimi.it/download/bzzmath-download/.

The second and larger part is a quick and effective overview of several frequently used non-conventional separation processes, which are mainly referred to heterogeneous phases and include concentration, filtration, drying, solid-fluid separation, and crystallization. Each chapter is independent from the others, except for some highlighted references, which are not essential for the understanding. Therefore, topics can be studied in individual chapters or by thematic groups.

The most used tool for the numerical solution of simple problem is Microsoft Excel, even if every other equivalent software can be suitable as well. The same remark holds true for the programming codes: they are written in C++, a widespread programming language, which is the most commonly used for process simulators. Nevertheless, the code transposition from C++ to any other programming language (e.g., Python, Fortran, Matlab, etc.) requires relatively poor efforts, so the same model equations can be solved by mean of every other language, given that a suitable ODEs and nonlinear equations system solver is available.

We wish finally to acknowledge the scientific and technical heritage bequeathed us by Prof. Sauro Pierucci, former professor of this course and endless source of inspiration. We would like to thank as well Prof. Guido Buzzi-Ferraris for all his teachings and nonetheless for his remarkable hard work in creating the BzzMath C++ library.

Milan, Italy   Alessandro Di Pretoro
                                                                                              Flavio Manenti

Contents

Nomenclature

Pinch Technology

c_P	Specific heat at constant pressure
F	Mass flowrate
$\dot{H}$	Enthalpy per unit time
$\dot{H}_{ref}$	Enthalpy per unit time
$\dot{Q}$	Heat per unit time
ΔT_{min}	Minimum temperature difference
T_{in}	Inlet temperature
T_{ref}	Reference temperature
T_{out}	Outlet temperature

Heat Pumps

A_C	Condenser heat transfer area
A_E	Evaporator heat transfer area
β	Compression ratio
c_P^L	Liquid specific heat at constant pressure
c_P^V	Vapor specific heat at constant pressure
$\Delta H_{ev}(T_i)$	Evaporation enthalpy at i-th temperature
L	Liquid flowrate
MW	Molecular weight
P_s	Vapor pressure
P_C	Condenser pressure
P_E	Evaporator pressure
Q_C	Heat provided in the condenser
Q_d	Dissipated heat
Q_E	Heat provided in the evaporator

R	Refrigerant flowrate
ΔT_{min}	Minimum temperature difference
T_C	Condenser temperature
T_E	Evaporator temperature
T_{cold}	Cold source temperature
T_{hot}	Hot sink temperature
U_C	Condenser heat transfer coefficient
U_E	Evaporator heat transfer coefficient
V	Vapor flowrate
W	Work

Cooling Towers

A	Heat exchanger surface area
a	Packing surface to height ratio
c_P	Specific heat at constant pressure
c_u	Humid air specific heat at constant pressure
G	Humid air flowrate
G_{dry}	Dry air flowrate
ΔH_{ev}	Heat of vaporization
H_G	Gas enthalpy
H_L	Liquid enthalpy
h_g	Gas side heat transfer coefficient
h_l	Liquid side heat transfer coefficient
J	Convective mass flow
K_U	Overall mass transfer coefficient
L	Liquid flowrate
P	Pressure
P_{ev}	Vapor pressure
Q	Heat exchanger duty
dS	Infinitesimal transfer area
T_a	Air temperature
T_c	Critical temperature
T_G	Gas temperature
T_L	Liquid temperature
ΔT_{min}	Minimum temperature difference
ΔT_{ml}	Logarithmic mean temperature difference
T_{wb}	Wet bulb temperature
U	Absolute humidity
U_s	Absolute humidity at saturation
W	Heat exchanger heat transfer coefficient
z	Spatial coordinate
Z	Relative humidity
dz	Infinitesimal height step

Multiple-Effect Evaporation

A_i	Heat exchanger surface area
$c_P^{solution}$	Solution specific heat at constant pressure
ΔH_{ev}^{steam}	Steam duty heat of vaporization
$\Delta H_{ev,i}^{vap}$	i-th effect vapor heat of vaporization
L_i	i-th effect liquid flowrate
P_i	i-th effect pressure
P_{ev}	Vapor pressure
T_i	i-th effect temperature
T_{v_i}	i-th vapor condensation temperature
$\Delta T_{eb,i}$	i-th effect boiling-point elevation
ΔT_i	i-th effect heat exchanger temperature difference
U	Heat exchanger heat transfer coefficient
V_i	i-th vapor flowrate
x_i	i-th solute concentration

Filtration

A	Filtration area
a	Filtrate fraction
C_s	Solid concentration
D	Channel diameter
D_P	Particle diameter
f	Friction factor
H	Basket height
K_i	Darcy's constant
L_0	Filter thickness
ΔL	Cake thickness
m_{cake}	Cake mass
n	Daily cycles
P	Pressure
ΔP	Pressure drops
Q	Flowrate
R	Outer radius
R_i	Inner radius
Re	Reynolds number
r	Radial coordinate
S_l	Lateral surface
t_{cycle}	Cycle duration
t_D	Decantation time
t_{down}	Downtime

t_F	Filtration time
t_W	Washing time
V_0, V_{loaded}	Loaded solution volume
V_F	Filtrate volume
V_W	Washing volume
v	Linear velocity
α	Filtrate vs. washing volume ratio
ε	Cake void fraction
μ	Viscosity
ξ	Loaded fraction
ρ_{cake}	Cake density
ρ_s	Solid density
ρ_w	Water density
ω	Angular velocity

Drying

A	Slice cross sectional area
c_P	Bread specific heat at constant pressure
k	Bread thermal conductivity
h	Heat transfer coefficient
H	Slice thickness
ΔH_{ev}	Water vaporization enthalpy
L	Slice side
m_w	Water content
P	Pressure
t	Time
T_0	Initial temperature
T_i	i-th subslice temperature
T_g	Gas temperature
W_0	Initial moisture
ρ_s	Bread density

Spray Drying

A	Spray dryer cross sectional area
c_P^g	Gas specific heat capacity at constant pressure
c_P^m	Particle specific heat capacity at constant pressure
C	Concentration
D	Equipment diameter

D_P Particle diameter
D_{va} Vapor diffusivity in air
f Friction factor
fat Fat percentage
g Gravitational acceleration
G Gas flowrate
G_{dry} Dry air flowrate
h Heat transfer coefficient
ΔH_{ev} Heat of vaporization
k Thermal conductivity
K_C Mass transfer coefficient
K_P Mass transfer coefficient
m_P Particle mass
m_{dry} Dry particle mass
MW Molecular weight
Nu Nusselt number
P Pressure
P^0 Vapor pressure
P_w Water partial pressure
Pr Prandtl number
Q_{fat} Fat flowrate
Q_l Liquid flowrate
Q_{milk} Milk flowrate
R Gas constant
Re Reynolds number
S_P Particle external surface
Sc Schmidt number
Sh Sherwood number
t Residence time
T_g Gas temperature
T_P Particle temperature
v_g Gas absolute velocity
v_P Particle absolute velocity
v_s Particle relative velocity
V_P Particle volume
W Moisture
z Vertical coordinate
η Number of particles per unit volume
μ Viscosity
ρ_g Gas density
ρ_P Particle density

Lyophilization

A	Cross sectional area
h	Heat transfer coefficient
ΔH_s	Sublimation enthalpy
J_C	Conductive heat flux
J_I	Radiant heat flux
J_T	Global heat flux
k	Solid thermal conductivity
t	Residence time
T_F	Sublimation temperature
T_P	Plate temperature
T_S	Surface temperature
T_S^{max}	Maximum allowed surface temperature
U	Global heat transfer coefficient
W	Moisture
x	Sublimation front coordinate
ρ_s	Solid density
σ	Radiant heat transfer coefficient

Gas-Solid Separation

A	Cyclone height
B	Cyclone inlet height
B	Abatement chamber width
D_P	Particle diameter
g	Gravitational acceleration
H	Cyclone inlet width
H	Abatement chamber height
N	Cyclone revolutions
Q	Gas flowrate
R	Cyclone radius
r	Radial coordinate
V_C	Linear velocity
V_D	Drift velocity
η	Separation efficiency
μ	Gas viscosity
ρ_g	Gas density
ρ_s	Solid density

Centrifugal Sedimentation

B	Circumference length
D_P	Particle diameter
h	Distance between two discs
H	Vertical distance between two discs
K_C	Convective velocity constant
K_D	Drift velocity constant
K_D'	Drift velocity constant
n	Number of discs
Q	Feed flowrate
V_C	Linear velocity
V_D	Drift velocity
x	Tangential coordinate
y	Normal coordinate
φ	Disc angle
μ	Gas viscosity
ρ_l	Liquid density
ρ_s	Solid density
ω	Angular velocity

Centrifugal Decantation

d	Outlet diameter
P	Pressure
Q	Volumetric flowrate
Q^w	Mass flowrate
r	Radial coordinate
R	Liquid-liquid interface radius
R_1	Light liquid interface radius
R_2	Heavy liquid interface radius
R_t	Decanter total radius
S	Outlet cross section
v	Outlet velocity
x	Mass fraction
ρ	Density
ω	Angular velocity

Membrane Separation

C_i	Concentration
F_i	Molar flowrate
J_i	Diffusive flux
N	Number of fibers per tubular module
n_i	Number of moles
P_i	Partial pressure
$Perm_i$	Permeability
r	Hollow fiber radius
Ret	Retention
dS	Infinitesimal surface area
T	Temperature
V	Volume
x_i	Molar fraction
z	Axial coordinate
$\alpha_{i,j}$	Separation factor

Crystallization

C_B	Bulk concentration
C_i	Interface concentration
C_S	Crystal equilibrium concentration
G_R	Growth ratio
J_r	Reactive mass flow
K	Overall mass transfer coefficient
K_C	Mass transfer coefficient
k_R	Reaction constant
L	Characteristic linear dimension
$\bar{L}_i^0$	i-th mesh average initial size
$\bar{L}_i^f$	i-th mesh average final size
ΔL	Growth (linear)
m_C	Single crystal mass
N	Reaction order
S_C	Single crystal external surface
t	Time
V_i^0	i-th mesh initial volume
V_i^f	i-th mesh final volume
W_i^0	i-th mesh initial weight
W_i^f	i-th mesh final weight
ρ_s	Crystal density

ϕ Volumetric shape factor

ψ Surface shape factor

ω_i^0 i-th mesh initial mass fraction

Part I
Heating and Cooling Operations

Chapter 1
Pinch Technology

Abstract Pinch technology is the best practice for heat exchanger networks optimization from an OPerating EXpenses point of view. Its main goal is to find a compromise between the investment required for the additional units and the external duty demand. The standard procedure is based on a preliminary thermodynamic assessment followed by the network optimization according to the feasibility of the operation. The results thus obtained allow then the decision maker to select the optimal solution according the specific needs of the case under analysis.

Keywords Energy · Heat exchanger network · Pinch · OPEX · Optimization

Pinch technology is a methodology aimed to optimize a process scheme from an energetical point of view, i.e. to define the optimal heat exchanger network configuration in order to minimize the external duty requirement. In general external duties cost decreases by approaching the ambient temperature as shown in Fig. 1.1, therefore the fundamental idea this method is based on is to take advantage of the heat exchange at the higher cost of the energy. Capital costs, proportional to the heat exchanger areas, are limited by fixing a ΔT_{min}, that will determine the so called "pinch condition" or "pinch point".

The optimal network resulting from the analysis usually includes a big number of heat exchangers. This means that whether the operating cost is minimized the total cost is not, therefore an a posteriori analysis should take into account all the cost items.

Further improvements can be obtained by "shifting" some streams or coupling small exchangers involving the same streams.

Since pinch technology was introduced the network design is performed following the analysis only if it shows an optimal condition far from the real one.

Good rules of thumb are:

- Do not transfer heat across the pinch point;
- Add heat above the pinch point;
- Remove heat below the pinch point;
- Add heat at the higher possible temperature with respect to the pinch point;
- Remove heat at the lower possible temperature with respect to the pinch point.

© The Author(s), under exclusive license to Springer Nature Switzerland AG 2020
A. Di Pretoro and F. Manenti, *Non-conventional Unit Operations*,
PoliMI SpringerBriefs, https://doi.org/10.1007/978-3-030-34572-3_1

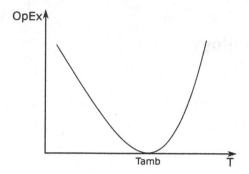

Fig. 1.1 Heating cost

1.1   Optimization of a Heat Exchanger Network

The minimum energy consumption condition of the process scheme shown in Fig. 1.2 needs to be evaluated. Streams data are listed in Table 1.1. Moreover, the suggestion of an alternative heat exchanger network is requested.

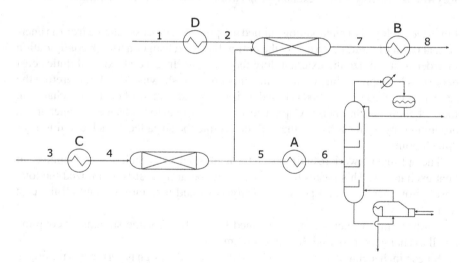

Fig. 1.2 Process flowsheet diagram

Table 1.1 Chemico-physical properties

Stream	1	2	3	4	5	6	7	8
Heat capacity (MW/°C)	0.3	0.3	0.2	0.2	0.15	0.15	0.25	0.25
T (°C)	160	250	40	200	270	60	220	100

Solution

First of all a few observations about pinch technology fundamentals are worth to be done. It is quite straightforward noticing that the two conditions this methodology imposes can be associated to the thermodynamics laws. Indeed the assessment of the minimum duty needed to achieve specifications is nothing but an overall energy balance, it takes into account how many Joules the hot streams can transfer to the cold ones and quantifies the difference; anyway it tells nothing about how to do that. On the other hand the ΔT_{min} condition reflects not only an economic constraint but also the need of a driving force for a feasible heat transfer. That is, it relates the energy quantity to the energy quality: 10 kJ are not the same at 50 °C or at 1000 °C.

The very first step of the solution procedure for the first request is the classification between hot and cold streams, where "hot" means that they need to be cooled down and "cold" means that they need to be heated up. The heat each stream should exchange can be then calculated as:

$$\dot{Q} = F \cdot c_P \cdot (T_{out} - T_{in}) \tag{1.1.1}$$

Table 1.2 resumes streams properties. According to the first law of thermodynamics it is straightforward to notice that a minimum cool duty of 2.5 MW is required to achieve the desired specifications.

However, this is not the final answer to this problem since it doesn't take into account the heat transfer feasibility. The second step of pinch analysis is then the outline of the so called "composite curve", consisting of an enthalpy-based streams classification and its representation in a T versus H diagram. Since enthalpy is a thermodynamic potential, only enthalpy differences with respect to a reference state can be measured. The selected reference state corresponds to the lowest among the hot streams ones, i.e. 60 °C. Since pressure can be considered constant and no phase transitions occur, the hot streams enthalpic states can be easily calculated as:

$$\dot{H}_2 = \dot{H}_1 + F \cdot c_P \cdot (T_2 - T_1) \tag{1.1.2}$$

Table 1.2 Exchanged heat versus heat exchanger

	Heat exchanger	T_{in} (°C)	T_{out} (°C)	ΔT (°C)	$F \cdot c_P$ (MW/°C)	$\dot{Q}$ (MW)
Hot	A	270	60	−210	0.15	−31.5
	B	220	100	−120	0.25	−30
Cold	C	40	200	160	0.2	32
	D	160	250	90	0.3	27

Table 1.3 Cascade diagram

HOT				COLD			
T (°C)	Stream	$F \cdot c_P$	$\dot{H}$ (MW)	T (°C)	Stream	$F \cdot c_P$	$\dot{H}$ (MW)
60	A	0.15	0	40	C	0.2	4
100	A	0.15	6	60	C	0.2	8
220	A+B	0.4	54	160	C	0.2	28
270	A	0.15	61.5	200	C+D	0.5	48
				250	D	0.3	63

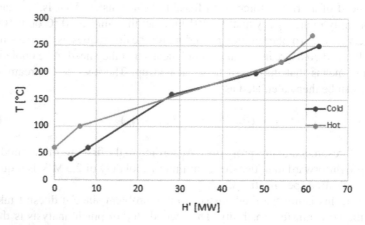

Fig. 1.3 Composite curve

The same formula applies to cold streams. Anyway in this case the enthalpy value referred to the lowest temperature, i.e. 40 °C, is evaluated as heat to be removed, that is the opposite of the enthalpy difference with respect to the reference state:

$$\dot{H}_{40\,°C} = \dot{H}_{ref} + F_C \cdot c_{P_C} \cdot (T_{ref} - T_C) = 4\,MW \qquad (1.1.3)$$

Enthalpy values obtained this way are then listed in Table 1.3. The corresponding plot, called "composite curve", is shown in Fig. 1.3. The $\Delta T_{min} = 10\,°C$ condition is clearly not satisfied in correspondence of more than one point of the chart. Moreover, there are some points where cold streams have higher temperature than hot ones.

In order to know the minimum enthalpic value (and the corresponding duty) required to satisfy the pinch condition the heat exchanger network needs to be solved starting from the lowest temperature. The first matches are between the cold thermal duty and the hot stream A and, after 100 °C, A+B; then the third match occurs between A+B and C. After that, by mean of the Microsoft Excel "Goal seek" function, the duty value $\dot{Q}_c = 10\,MW$ corresponding to $\Delta T_{min} = 10\,°C$ can be easily found. The new composite curve data and its respective plot are reported in Table 1.4 and Fig. 1.4.

Table 1.4 Cascade diagram – corrected

HOT				COLD			
T (°C)	Stream	$F \cdot c_P$	$\dot{H}$ (MW)	T (°C)	Stream	$F \cdot c_P$	$\dot{H}$ (MW)
60	A	0.15	0	40	C	0.2	10
100	A	0.15	6	60	C	0.2	14
220	A+B	0.4	54	160	C	0.2	34
270	A	0.15	61.5	200	C+D	0.5	54
				250	D	0.3	69

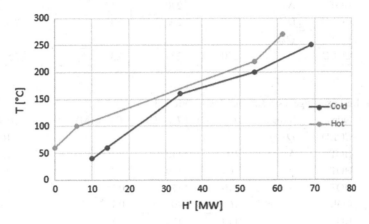

Fig. 1.4 Composite curve – corrected

Actually it can be already stated, according to the first thermodynamic law, that:

$$\dot{Q}_{hot} = \dot{Q}_{cold} - 2.5 = 7.5 \, \text{MW} \tag{1.1.4}$$

However, it is worth keep on matching cold and hot streams in order to accurately define the whole heat exchanger network. Results are listed in Table 1.5.

We can then finally answer to the first question of the problem stating that a minimum of 10 MW cold duty and 7.5 MW hot duty are needed to achieve the required specifications.

The answer to the second request is not straightforward even if the streams matches corresponding to each temperature range are already known. Streams indeed cannot be mixed as suggested for the exchangers 2, 3 and 4. It means these streams need to be split between the corresponding temperatures and more heat exchangers should be used. One of the possible real heat exchanger networks is given by the matches listed in Table 1.6. The corresponding PFD is reported in Fig. 1.5.

As already mentioned in the introduction, pinch technology is applied only if operating costs reduction justifies the increase in heat exchanger units and their respective maintenance costs. Figure 1.5 shows that 9 heat exchangers are required instead of

Table 1.5 Heat exchangers streams matches

#	Side	Stream	T_{out} (°C)	T_{in} (°C)	$F \cdot c_P$ (MW/°C)	Q (MW)
	COLD	Q_{cold}				−10
1	HOT	A	60	100	0.15	−6
2	HOT	A+B	100	110	0.4	−4
3	HOT	A+B	110	170	0.4	−24
	COLD	C	160	40	0.2	24
4	HOT	A+B	170	220	0.4	−20
	COLD	C+D	200	160	0.5	20
5	HOT	A	220	270	0.15	−7.5
	COLD	D	225	200	0.3	7.5
	HOT	Q_{hot}				7.5
6	COLD	D	250	225	0.3	7.5

Table 1.6 Heat exchangers streams matches – corrected

#	Side	Stream	T_{out} (°C)	T_{in} (°C)	$F \cdot c_P$ (MW/°C)	Q (MW)
	COLD	Q_{cold}				−10
1	HOT	A	60	110	0.15	−7.5
2	HOT	B	100	110	0.3	−2.5
3.1	HOT	B	110	170	0.25	−15
	COLD	0.625*C	160	40	0.125	−15
3.2	HOT	A	110	170	0.15	−9
	COLD	0.375*C	160	40	0.075	9
4.1	HOT	A	170	220	0.15	−7.5
	COLD	0.9375*C	200	160	0.1875	7.5
4.2	HOT	0.96*B	170	220	0.24	−12
	COLD	D	200	160	0.3	12
4.3	HOT	0.04*B	170	220	0.01	−0.5
	COLD	0.0625*C	200	160	0.0125	0.5
5	HOT	A	220	270	0.15	−7.5
	COLD	D	225	200	0.3	7.5
	HOT	Q_{hot}				7.5
6	COLD	D	250	225	0.3	7.5

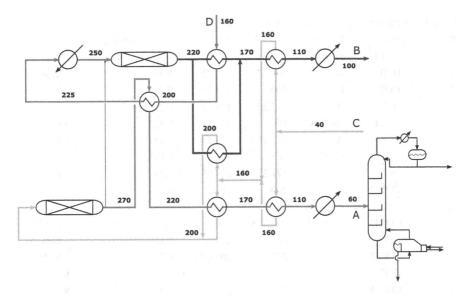

Fig. 1.5 Heat exchanger network

the 4 ones reported in Fig. 1.2 to obtain a decrease of external duty consumption from 61.5 and 59 MW to 10 and 7.5 MW. After an a posteriori economic analysis a compromise between number of units and external duty requirement could be found and an intermediate configuration should be selected.

For instance through the exchanger 4.3 only 0.5 MW are transferred; this is the typical case when it's worth paying an additional external duty and removing such a small unit. We can then decide to increase both the cold and hot duties by 0.5 MW and obtain a new heat exchanger network that doesn't satisfy the minimum operating costs condition but results in a lower total cost. The new matches are listed in Table 1.7 and the corresponding PFD is reported in Fig. 1.6.

1.2 Conclusions

This chapter aims to show how powerful pinch technology is from an energy efficiency point of view. Although being a complex methodology that depends on the specific heat exchange network, it is based on considerations of general validity deriving from the thermodynamic laws. Its main features, widely explored through an application case study, can be adapted to every system. The effort required for the optimization of a rather simple problem suggests, however, the difficulty to cope with in case of a chemical plant section to analyze.

Table 1.7 Heat exchangers streams matches – final

#	Side	Stream	T_{out} (°C)	T_{in} (°C)	$F \cdot c_P$ (MW/°C)	Q (MW)
	COLD	Q_{cold}				−10.5
1	HOT	A	60	111.25	0.15	−7.6875
2	HOT	B	100	111.25	0.25	−2.8125
3.1	HOT	B	111.25	171.25	0.25	−15
	COLD	0.625*C	160	40	0.125	15
3.2	HOT	A	111.25	171.25	0.15	−9
	COLD	0.375*C	160	40	0.075	9
4.1	HOT	A	171.25	224.58	0.15	−8
	COLD	C	200	160	0.2	8
4.2	HOT	B	171.25	220	0.25	−12.1875
	COLD	D	200.625	160	0.3	12.1875
5	HOT	A	224.58	270	0.15	−6.8125
	COLD	D	223.33	200.625	0.3	6.8125
	HOT	Q_{hot}				8
6	COLD	D	250	223.33	0.3	8

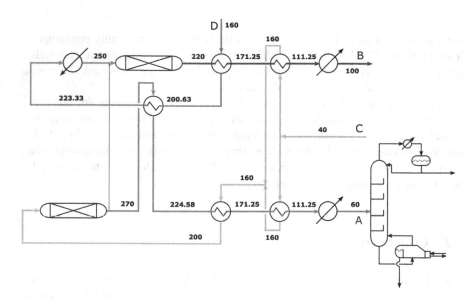

Fig. 1.6 Heat exchanger network – final

Indeed, pinch technology is mostly used to cut the costs deriving from considerable external heat duty requirement and it is seldom applied in a rigorous way. Even when a stricter enforcement is desired, given the high number of heat exchanger needed, the resulting optimal design is always subjected to an economic assessment and validated only if really profitable with respect to the standard configuration. At the end the usual design choice is usually a compromise between the optimal and the actual ones as shown in the example.

Chapter 2
Heat Pumps

Abstract Heat pumps are widespread units employed in the daily life whether a consistent amount of energy is available at a low enthalpic level (i.e. not suitable temperature). Their particular efficiency is shown by mean of a practical example where heat balances and operating conditions are thoroughly discussed. The sizing of each unit (evaporator, condenser and compressor) follows the process design in order to provide a complete overview.

Keywords Refrigeration cycle · Vapor compression · Heating · COP

A heat pump is a device aimed to transfer heat from a low temperature source to a higher temperature heat sink. In order to move thermal energy in the opposite direction of the thermal gradient external power is required according to the Clausius statement[1] of the second law of thermodynamics (cf. Fig. 2.1).

In order to enhance the thermal level of the heat to be provided to the hot sink, the fluid (refrigerant) undergoes adiabatic compression (compression heat pumps); in case of a low cost or waste available thermal energy source compression is substituted by a more complex chemical/physical absorption system of a refrigerant in a carrier fluid (absorption heat pumps). The high efficiency of heat pumps is due to the fact that you don't have to produce the energy you need, you just require to spend the amount of energy necessary to move it.

Actually heat pump and refrigeration thermodynamic cycles are analogous, the difference lies in their purpose: while the former is aimed to provide heat to a sink whose temperature is higher than the ambient one, the latter has the purpose to remove heat from a source at a temperature lower than T_{amb}.

The common design of a compression heat pump involves four main components: a condenser, an expansion valve, an evaporator and a compressor as shown in Fig. 2.2. The coefficient of performance (COP), defined as the ratio between the useful energy and the energy to be paid, varies from 5–7 for big machines and 2–3 for small ones. In order to compare these values to the efficiency of the most common heating

---

[1]Clausius, R. (1854). "Über eine veränderte Form des zweiten Hauptsatzes der mechanischen Wärmetheorie". Annalen der Physik. Poggendoff. xciii: 481–506. https://doi.org/10.1002/andp. 18541691202.

Fig. 2.1 Heat pump thermal
machine scheme

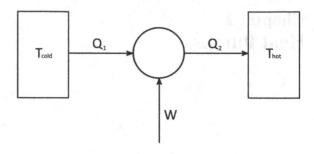

Fig. 2.2 Vapor compression
heat pump

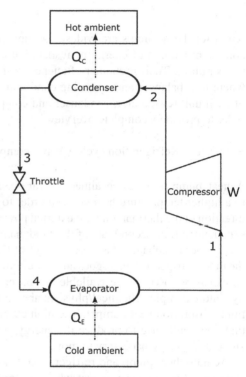

systems, e.g. boilers, the cost of electrical energy needs to be converted to thermal
energy basis. Electricity indeed costs about three times the same amount of thermal
calories, therefore the aforementioned COP values should be divided by 3. However,
being the COP of boilers approximately equal to 0.75, heat pumps still result to be
much more profitable.

Heat pumps allow to reduce energetical consumption, atmospheric pollution as
well as process heating costs. Moreover they're environmental friendly since they
use highly dispersed energy sources such as heat recovery from cooling water or
from cold air. Even if heat pumps result to be very effective and to allow high energy
savings, they're not a very spread technology due to the high capital costs required

that vanish their profitability and to the uncertain maintenance costs. Moreover their use could be limited by the high cost of electrical energy and the replacement of the fuel gas network by this technology is unfeasible due to the inability of the actual power plants and relative distribution network to withstand such a high load.

2.1 Heat Pump Design

A 7.5 kW vapor compression heat pump to heat up a 95 °C system should be designed. The ambient temperature is 50 °C and a minimum temperature difference $\Delta T_{min} = 5$ °C is required for a feasible exchange. The refrigerant is isopentane whose properties are listed in Table 2.1.

The Antoine coefficients correspond to the equation expressed as:

$$log_{10}(P_s[bar]) = A - \frac{B}{T[K] + C} \qquad (2.1.1)$$

Moreover, given $U_C = 375 \, \text{W}/(\text{m}^2 \cdot \text{K})$ and $U_E = 340 \, \text{W}/(\text{m}^2 \cdot \text{K})$, the equipment sizing is requested.

Solution I

Given the minimum temperature difference, condenser and evaporator operating temperatures can be easily evaluated as:

$$T_C = T_{hot} + \Delta T_{min} = 100 \, ^\circ C \qquad (2.1.2)$$

$$T_E = T_{cold} - \Delta T_{min} = 45 \, ^\circ C \qquad (2.1.3)$$

Table 2.1 Isopentane properties

	Property	Value	Unit
Antoine coefficients			
	A	3.97183	
	B	1021.864	
	C	−43.231	
Physical properties			
	c_P^L	2.426	kJ/(kg · K)
	c_P^V	1.778	kJ/(kg · K)
	$\Delta H_{ev}(T_C)$	19940	kJ/kmol
	MW	72.15	g/mol

Since phase transition occurs both in condenser and evaporator, pressure can be assessed by using the Antoine correlation:

$$P_C = P_s(T_C) = 10^{A - \frac{B}{T_C[K]+C}} \tag{2.1.4}$$

$$P_E = P_s(T_E) = 10^{A - \frac{B}{T_E[K]+C}} \tag{2.1.5}$$

The specification related to the power of the pump corresponds to the heat provided to the hot ambient, i.e. Q_C; once the condenser temperature is known, the corresponding evaporation heat can be found in literature. The required refrigerant flowrate results then to be:

$$R = \frac{Q_C}{\Delta H_{ev}(T_C)} = 0.027\,\text{kg/s} \tag{2.1.6}$$

On the other hand the heat required for fluid evaporation corresponds to:

$$Q_E = L \cdot \Delta H_{ev}(T_E) \tag{2.1.7}$$

However, nor L neither $\Delta H_{ev}(T_E)$ are known. L is the non-vaporized refrigerant flowrate after throttling whose value can be calculated by imposing mass and energy balances across the valve:

$$\begin{cases} R = L + V \\ R \cdot c_P^L \cdot T_C + Q_d = L \cdot c_P^L \cdot T_E + V \cdot (c_P^L \cdot T_E + \Delta H_{ev}(T_E)) \end{cases} \tag{2.1.8}$$

where Q_d is the heat dissipation during the expansion. This quantity is nevertheless negligible and the system can be solved as an adiabatic expansion:

$$\begin{cases} V = R - L \\ R \cdot c_P^L \cdot T_C = L \cdot c_P^L \cdot T_E + (R - L) \cdot (c_P^L \cdot T_E + \Delta H_{ev}(T_E)) \end{cases} \Longleftrightarrow$$

$$\begin{cases} V = R - L \\ L \cdot \Delta H_{ev}(T_E) = R \cdot (\Delta H_{ev}(T_E) - c_P^L \cdot (T_C - T_E)) \end{cases} \Longleftrightarrow$$

$$\begin{cases} V = R \cdot \frac{c_P^L \cdot (T_C - T_E)}{\Delta H_{ev}(T_E)} \\ L = R \cdot (1 - \frac{c_P^L \cdot (T_C - T_E)}{\Delta H_{ev}(T_E)}) \end{cases} \tag{2.1.9}$$

Therefore the previous equation can be rewritten as:

$$Q_E = R \cdot (\Delta H_{ev}(T_E) - c_P^L \cdot (T_C - T_E)) \tag{2.1.10}$$

where $\Delta H_{ev}(T_E)$ can be easily found in manuals. Anyway, if we want to refer calculations to the same temperature, let's say T_C, we can take advantage of the Kirchhoff equation:

$$\Delta H_{ev}(T_E) = \Delta H_{ev}(T_C) + (c_P^V - c_P^L) \cdot (T_E - T_C) \tag{2.1.11}$$

By substituting $\Delta H_{ev}(T_E)$ in the Q_E equation, the heat duty at the evaporator becomes:

$$Q_E = R \cdot (\Delta H_{ev}(T_C) - c_P^V \cdot (T_C - T_E)) = 4.85\,\text{kW} \tag{2.1.12}$$

and the work expression resulting from the overall energy balance can be simplified as:

$$W = Q_C - Q_E = R \cdot c_P^V \cdot (T_C - T_E) = 2.65\,\text{kW} \tag{2.1.13}$$

The COP is then given by the ratio between the useful energy and the energy to be paid, that is:

$$COP = \frac{Q_C}{W} = \frac{\cancel{R} \cdot \Delta H_{ev}(T_C)}{\cancel{R} \cdot c_P^V \cdot (T_C - T_E)} = 2.83 \tag{2.1.14}$$

It is worth noticing that the coefficient of performance doesn't depend on the amount of circulating refrigerant while, differently from ideal fluids (e.g. Carnot cycle), it depends on its nature since both ΔH_{ev} and c_P^V are present in the equation.

Solution II

Equipment sizing as well is not a cumbersome task. The expansion valve is sized according to the pipeline diameter and its opening is manipulated to control the pressure difference. Evaporator and condenser are sized by mean of the heat exchanger characteristic equation:

$$A_C = \frac{Q_C}{U_C \cdot \Delta T_{min}} = 4\,\text{m}^2 \tag{2.1.15}$$

$$A_E = \frac{Q_E}{U_E \cdot \Delta T_{min}} = 2.85\,\text{m}^2 \tag{2.1.16}$$

The temperature difference can be considered constant since phase transition is occurring in tube side and outside temperature doesn't change along the exchanger (not during time); for plate heat exchangers a lower ΔT_{min} can be selected (2–3 °C).

The remaining unit to be designed is the compressor; compressor selection occurs by referring to its characteristic curve that shows the relationship between flowrate and compression ratio. The latter is defined as the ratio between outlet and inlet pressure:

$$\beta = \frac{P_C}{P_E} = 4.17 \tag{2.1.17}$$

Hence, assuming constant operating conditions, the compressor has to be selected as the one with the peak efficiency line of the characteristic curve passing through the point (R, β). Actually operating conditions change according to external hot and cold temperatures fluctuations, therefore an inverter is connected to the compressor and manipulates the rotational speed (RPM) to adapt to the compression ratio variations.

2.2   Conclusions

The general procedure for the vapor compression heat pump design has been introduced in detail, being this thermodynamic cycle the most widespread in all common applications. Absorption heat pumps can be installed as well whether electricity is expensive, not easily accessible or if waste or cheap thermal energy is available.

The case study above shows that, considering an ideal cycle, the COP depends only on the source and sink temperatures as well as on the refrigerant nature, as expected. To perform the design under real conditions, both non-isentropic expansion and compression should be taken into account by mean of the isentropic efficiency parameter.

Chapter 3
Cooling Towers

Abstract Cooling towers are the most used units designed to reduce cooling water temperature downstream the process both in once-through configurations and in those with a recycle. Their theoretical background is strictly related to the definition of humidity and wet bulb temperature. From a modeling point of view they can be solved as a standard desorption column whose specification is the outlet liquid temperature; the corresponding BVP can be smartly turned into as a standard ODE system to converge making calculations easier and allowing for the design of a packed bed unit as a case study.

Keywords Humidity · Wet bulb · Desorption · Packed bed

Water results to be the most widespread utility in chemical plants since it is easily accessible, widely available and rather cheap (practically free whether no expensive pretreatments are required).
Water can enter the process plant in two forms for different purposes:

- Steam: as heating duty;
- Cool water: as cooling duty.

Cooling water temperature increases along the progress of the process until the downstream battery limit. Directly contacting hot water with dry air, cooling towers are aimed to reduce the water outlet temperature below the air temperature itself at the cost of a water loss by evaporation. This operation is performed both in case of recycle and in once-through processes respectively to reuse water as cooling utility or to discharge it in the environment according to the legislation constraints.

3.1 Cooling Tower Packing Design

A 20000 kg/h water stream at 50 °C, coming from a heat exchanger, should be cooled down to 25 °C as required by the law before being discharged. For this purpose a packed bed cooling tower, where direct contacting between water and air takes place, will be used. The dry air is collected from the external environment at 20 °C and 50% of relative humidity with a 15000 kg/h flowrate.

© The Author(s), under exclusive license to Springer Nature Switzerland AG 2020   19
A. Di Pretoro and F. Manenti, *Non-conventional Unit Operations*,
PoliMI SpringerBriefs, https://doi.org/10.1007/978-3-030-34572-3_3

Table 3.1 Chemico-physical properties

Property	Value	Unit
$< c_u >$	0.26	kcal/(kg · K)
c_P^L	1	kcal/(kg · K)
ΔH_{ev}	580	kcal/kg
$h_g \cdot a$	4000	kcal/(m · h · K)
$h_l \cdot a$	30000	kcal/(m · h · K)

The packing height and the air wet bulb calculations are requested.

The chemico-physical properties are listed in Table 3.1. The vapor pressure linearization in the range [25, 50] °C states as:

$$P_{ev}(mmHg) = -49.705 + 2.71 \cdot T(°C) \tag{3.1.1}$$

Solution I

Looking closely at this problem, the analogy of the cooling tower design with the design of a packed bed desorption column results evident. Actually the same identical solving procedure will be applied with the only difference that, in the desorption case, the design specification is in general the molar fraction of a liquid component while for cooling towers the design requirement is the outlet liquid temperature.

In order to better understand mass and heat balances we'll refer to Fig. 3.1. The interphase mass flow can be written on a humidity basis as:

Fig. 3.1 Mass and heat transfer scheme

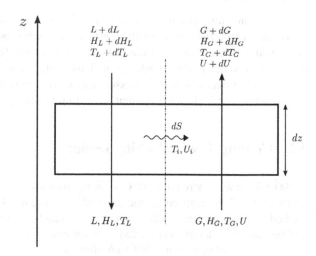

$$dJ = K_U \cdot (U_i - U) \cdot dS \qquad (3.1.2)$$

The mass and heat balances for each phase can then be written as follows.

- Gas phase:

$$\begin{cases} dG = dJ = K_U \cdot (U_i - U) \cdot dS \\ d[G \cdot H_G] = G \cdot dH_G + dG \cdot H_G = h_G \cdot (T_i - T_G) \cdot dS + K_U \cdot (U_i - U) \cdot dS \cdot H_{G_i} \end{cases} \Longleftrightarrow$$

$$\begin{cases} dG = dJ = K_U \cdot (U_i - U) \cdot dS \\ G \cdot c_P^V \cdot dT_G + \cancel{K_U \cdot (U_i - U) \cdot dS} \cdot H_G = h_G \cdot (T_i - T_G) \cdot dS + \cancel{K_U \cdot (U_i - U) \cdot dS} \cdot H_{G_i} \end{cases}$$

Making the assumption that $H_G \cong H_{G_i}$ the terms on the left and right hand side of the equality can be simplified as:

$$\begin{cases} dG = K_U \cdot (U_i - U) \cdot dS \\ G \cdot c_P^V \cdot dT_G = h_G \cdot (T_i - T_G) \cdot dS \end{cases} \qquad (3.1.3)$$

Moreover, given the relationship:

$$G = G_{dry} \cdot (1 + U) \Longleftrightarrow dG = G_{dry} \cdot dU \qquad (3.1.4)$$

and

$$G \cdot C_P^V = G_{dry} \cdot (1 + U) \cdot c_P^V = G_{dry} \cdot c_u \qquad (3.1.5)$$

the final expressions are obtained:

$$\begin{cases} G_{dry} \cdot dU = K_U \cdot (U_i - U) \cdot dS \\ G_{dry} \cdot c_u \cdot dT_G = h_G \cdot (T_i - T_G) \cdot dS \end{cases} \qquad (3.1.6)$$

- Liquid phase

By analogy the liquid phase balances state as:

$$\begin{cases} -dL = -dJ = -K_U \cdot (U_i - U) \cdot dS \\ -d[L \cdot H_L] = -L \cdot dH_L - dL \cdot H_L = h_L \cdot (T_i - T_L) \cdot dS - K_U \cdot (U_i - U) \cdot dS \cdot H_{L_i} \end{cases} \Longleftrightarrow$$

$$\begin{cases} dL = K_U \cdot (U_i - U) \cdot dS \\ -L \cdot c_P^L \cdot dT_L - \cancel{K_U \cdot (U_i - U) \cdot dS} \cdot H_L = h_L \cdot (T_i - T_L) \cdot dS - \cancel{K_U \cdot (U_i - U) \cdot dS} \cdot H_{L_i} \end{cases}$$

Making the assumption that $H_L \cong H_{L_i}$ the terms on the left and right hand side of the equality can be simplified as:

$$\begin{cases} dL = K_U \cdot (U_i - U) \cdot dS \\ L \cdot c_P^L \cdot dT_L = h_L \cdot (T_L - T_i) \cdot dS \end{cases} \qquad (3.1.7)$$

These equations could be easily solved if the interface temperature T_i was known.

Before going any further in the resolution, a few terms that are still unkown need to be calculated.

Given the air relative humidity and temperature, the absolute humidity can be easily estimated as:

$$Z = \frac{U}{U_s(T_a)} \Longleftrightarrow U = Z \cdot 0.62 \cdot \frac{P_{ev}(T_a)}{P} \qquad (3.1.8)$$

The mass transfer coefficient can be evaluated by using the Lewis relationship:

$$\frac{h_G}{K_u \cdot c_u} = 1 \Longleftrightarrow K_u = \frac{h_G}{c_u} \qquad (3.1.9)$$

The heat transfer coefficients are multiplied by the term a. This term is defined as the transfer area per unit length of the packing, i.e.:

$$a = \frac{dS}{dz} \Longleftrightarrow dS = a \cdot dz \qquad (3.1.10)$$

All the required parameters to calculate the interface temperature are now available. In order to do that the heat balance on the adiabatic system needs to be solved:

$$d[G \cdot H_G] - d[L \cdot H_L] = 0 \Longleftrightarrow$$
$$\Longleftrightarrow h_G \cdot (T_i - T_G) \cdot dS + K_U \cdot (U_i - U) \cdot dS \cdot H_{G_i} + h_L \cdot (T_i - T_L) \cdot dS$$
$$- K_U \cdot (U_i - U) \cdot dS \cdot H_{L_i} = 0$$
$$\Longleftrightarrow h_L \cdot (T_L - T_i) \cdot a \cdot d\!\!\!/z = h_G \cdot (T_i - T_G) \cdot a \cdot d\!\!\!/z + K_U \cdot (U_i - U) \cdot a \cdot d\!\!\!/z \cdot (H_{G_i} - H_{L_I})$$
$$\Longleftrightarrow h_L \cdot a \cdot (T_L - T_i) = h_G \cdot a \cdot (T_i - T_G) + K_U \cdot a \cdot (U_i - U) \cdot \Delta H_{ev_i} \qquad (3.1.11)$$

Thanks to the hypothesis of constant vaporization heat and to the linearization of the vapor pressure correlation, the derived equation is linear with respect to T_i.

$$h_L \cdot a \cdot (T_L - T_i) = h_G \cdot a \cdot (T_i - T_G) + K_U \cdot a \cdot \left(0.62 \cdot \frac{A + B \cdot T_i}{P} - U\right) \cdot \Delta H_{ev_i} \quad (3.1.12)$$

Therefore T_i can finally be made explicit:

$$T_i = \frac{h_L \cdot a \cdot T_L + h_G \cdot a \cdot T_G + K_U \cdot a \cdot (U - 0.62 \cdot \frac{A}{P}) \cdot \Delta H_{ev}}{h_L \cdot a + h_G \cdot a + K_U \cdot a \cdot 0.62 \cdot \frac{B}{P} \cdot \Delta H_{ev}} \qquad (3.1.13)$$

The main issue during the integration of this ODE system is related to the initial conditions. Indeed both liquid and air temperature, air humidity and air flowrate at the outlet of the cooling tower are known while the value of the liquid flowrate is given at the inlet only. Apparently we are requested to solve a so called Boundary Value Problem but actually we can deal with it without any cumbersome procedure.

Table 3.2 Integration results

z	U	T_L	T_G	L	G
0.00	0.00183	25.00	20.00	19292	15000
0.20	0.00335	25.79	20.53	19315	15023
0.40	0.00489	26.60	21.10	19338	15046
0.60	0.00645	27.42	21.71	19361	15069
0.80	0.00804	28.26	22.34	19385	15093
1.00	0.00966	29.12	23.01	19409	15117
1.20	0.01131	30.00	23.70	19434	15142
1.40	0.01299	30.90	24.41	19459	15167
1.60	0.01470	31.81	25.15	19485	15193
1.80	0.01645	32.74	25.91	19511	15219
2.00	0.01824	33.70	26.70	19538	15246
2.20	0.02006	34.67	27.50	19565	15273
2.40	0.02192	35.66	28.33	19593	15301
2.60	0.02381	36.67	29.17	19622	15330
2.80	0.02575	37.69	30.03	19651	15359
3.00	0.02772	38.74	30.92	19680	15388
3.20	0.02973	39.81	31.82	19710	15418
3.40	0.03179	40.90	32.74	19741	15449
3.60	0.03388	42.01	33.69	19773	15481
3.80	0.03601	43.13	34.65	19805	15513
4.00	0.03819	44.28	35.63	19837	15545
4.20	0.04040	45.45	36.63	19871	15579
4.40	0.04266	46.64	37.65	19904	15612
4.60	0.04496	47.84	38.69	19939	15647
4.80	0.04730	49.07	39.74	19974	15682
4.95	0.04909	50.01	40.55	20001	15709

The value of outlet liquid flowrate (obviously lower than 20000 kg/h) can be assumed as known and the integration of the ODE system can be run. Specification is achieved for $T_L = T_w = 50\,°C$, the corresponding inlet liquid flowrate value is then compared to the given one and the outlet liquid flowrate assumption is modified accordingly. When both $T_L = 50\,°C$ and $L = 20000\,kg/h$ are attained at the same integration step, the ODE system equivalent to the BVP is solved and the corresponding value of z represents the height of the packing required to perform the cooling process as requested.

The results of the integration are shown in Table 3.2. The gas and liquid temperature as well as flowrate profiles are plotted respectively in Figs. 3.2 and 3.3.

The required height of the packed bed is $4.95\,m$ and the corresponding outlet liquid flowrate is $L = 19292\,kg/h$.

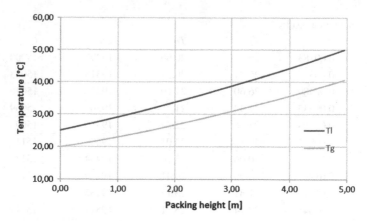

Fig. 3.2 Gas and liquid temperature profiles

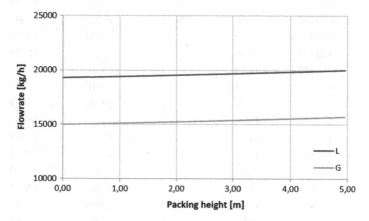

Fig. 3.3 Gas and liquid flowrate profiles

Solution II

Since the wet bulb temperature should be lower than the dry bulb one, i.e. 20 °C, the P_{ev} linearization cannot be used anymore given that it is valid in the range [25, 50] °C only.

The complete vapor pressure correlation[1] in the range [0, 200] °C is then proposed:

$$ln(P_{ev}(Pa)) = \frac{C_1}{T(K)} + C_2 + C_3 \cdot T + C_4 \cdot T^2 + C_5 \cdot T^3 + C_6 \cdot ln(T) \quad (3.1.14)$$

where the C_i constants are those reported in Table 3.3.

---

[1] ASHRAE Handbook—Fundamentals 2002, Chap. 6.

Table 3.3 Vapor pressure correlation coefficients

C_1	$-5.8002206 \cdot 10^3$
C_2	1.3914993
C_3	$-4.8640239 \cdot 10^{-2}$
C_4	$4.1764768 \cdot 10^{-5}$
C_5	$-1.4452093 \cdot 10^{-8}$
C_6	6.5459673

Table 3.4 Latent heat correlation coefficients

B_1	$-5.8002206 \cdot 10^3$
B_2	1.3914993
B_3	$-4.8640239 \cdot 10^{-2}$
B_4	$4.1764768 \cdot 10^{-5}$

In order to have a more accurate result it can be also taken into account the latent heat variation with temperature without any appreciable computational effort increase.

The water heat of vaporization correlation[2] in the range [273.15, 647.13] K states as:

$$dH_{ev}(\text{J/kmol}) = B_1 \cdot (1 - T_r)^{B_2 + B_3 \cdot T_r + B_4 \cdot T_r^2} \qquad (3.1.15)$$

where T_r is the reduced temperature ($T_{C,water} = 647.13$ K) and the B_i constants are those reported in Table 3.4.

The so called "wet bulb equation", representing the steady state heat balance, states as:

$$c_u \cdot (T_a - T_{wb}) + (U - U_s(T_{wb})) \cdot \Delta H_{ev}(T_{wb}) = 0 \qquad (3.1.16)$$

Despite the non-linearity of the equation it could, be easily solved with the Microsoft Excel "Goal seek" function.

The resulting value is $T_{wb} = 14.02\,°C$.

Furthermore, a second methodology to answer the question is proposed by using the psychrometric chart (cf. Fig. 3.4).

The wet bulb temperature depends on the air conditions only. In this case the air conditions are defined by dry-bulb temperature, relative humidity and pressure.

This means that the provided data allow the identification of the missing property, i.e. wet bulb temperature, in a few steps:

1. Pick up the chart referring to atmospheric pressure;
2. Find the relative humidity curve according to the given value (i.e. 50%);

---

[2]Perry's Chemical Engineers' Handbook 8th edition, table 2–150.

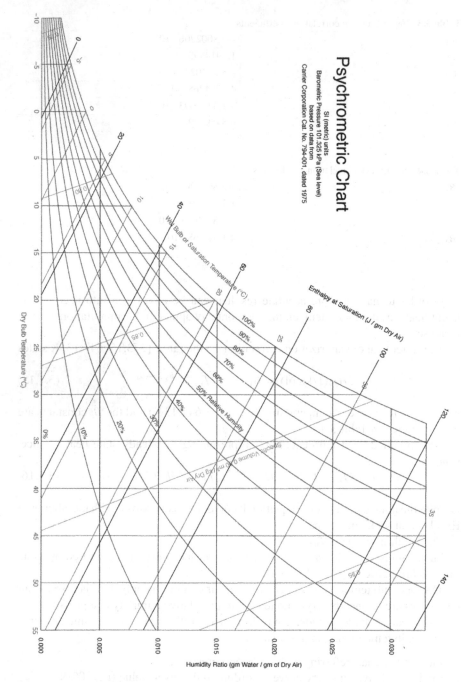

Fig. 3.4 Psychrometric chart (Sea level)

3. Find on the x-axis the dry bulb temperature value (i.e. 20 °C). Then draw a vertical line until crossing the relative humidity curve in the point that represents the physical condition of the air (yellow diamond in Fig. 3.5);
4. Starting from the aforementioned point draw a line with a slope of $-\frac{c_u}{\Delta H_{ev}}$ (i.e. parallel to the light blue ones) up to the intersection with saturation curve corresponding to the wet bulb temperature value (red point in Fig. 3.5).

The resulting wet bulb temperature is about 13.5 °C, value that perfectly agrees with the one obtained by the first procedure.

3.2 Additional Remarks

An additional evaluation worth to be done is the estimation of this cooling process cost in term of water losses.

In order to do this we can refer to the liquid phase mass balance:

$$- dL = -dJ = -K_U \cdot (U_i - U) \cdot dS \iff dL = K_U \cdot a \cdot (U_i - U) \cdot dz$$
$$(3.2.1)$$

Given the initial condition $L = 20000$ kg/h corresponding to $z = 4.95$ m, it can be integrated backwards until $z = 0$ obtaining the liquid flowrate profile already shown in Fig. 3.3.

The remaining cool water flowrate, related to the required water integration in case of recycle, is about 19292 kg/h. Therefore 708 kg/h, i.e. the 3.54 % of the overall liquid flowrate, are loss in order to reduce the water temperature by 25 °C.

Such a water loss corresponds to an enthalpy loss of 410640 kcal/h able to reduce by about 21.31 °C the remaining water flow temperature. This means that 21.3 °C out of 25 °C total temperature decrease have been obtained because of water evaporation.

In order to better understand the impact of phase transition in a cooling operation a heat exchanger performing the same operation without direct contacting between the two phases can be finally designed.

The required design data are reported in Table 3.5; a 15 °C air temperature will be considered in order to satisfy the ΔT_{min} constraint.

The demanded heat duty is:

$$Q = L \cdot c_P^{water} (T_{in,water} - T_{out,water}) = 482300 \, \text{kcal/h} \qquad (3.2.2)$$

The air outlet temperature and then the air flowrate are given by:

$$T_{out,air} = T_{in,water} - \Delta T_{min} = 40 \,°C \qquad (3.2.3)$$

$$L \cdot c_P^{water} \cdot \Delta T_{water} = G \cdot c_P^{air} \cdot \Delta T_{air} \iff G = \frac{L \cdot c_P^{water} \cdot \Delta T_{water}}{c_P^{air} \cdot \Delta T_{air}} = 81026 \, \text{kg/h}$$
$$(3.2.4)$$

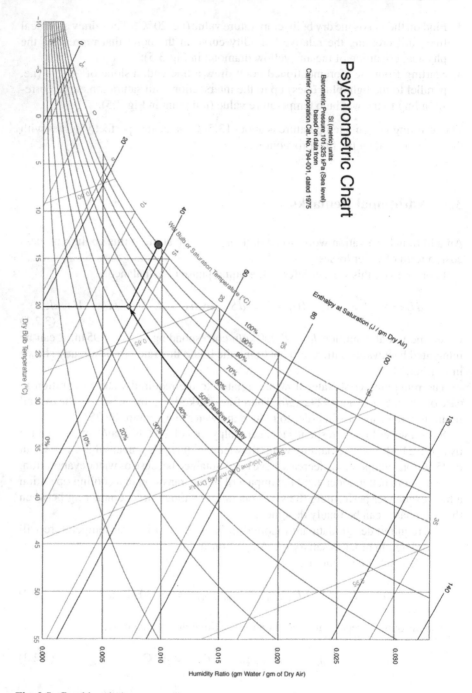

Fig. 3.5 Graphic solution

Table 3.5 Heat exchanger design data

Parameter	Value	Unit
L	19292	kg/h
$T_{in,water}$	50	°C
$T_{out,water}$	25	°C
$T_{in,air}$	15	°C
c_P^{water}/c_P^{air}	4.2	1
ΔT_{min}	10	°C
W	11.2[a]	kcal/h

[a] https://www.engineeringtoolbox.com/overall-heat-transfer-coefficients-d_284.html

Finally the heat exchanger area can be calculated using the equation:

$$Q = W \cdot A \cdot \Delta T_{ml} \iff A = \frac{Q}{W \cdot \Delta T_{ml}} = 4306 \, \text{m}^2 \qquad (3.2.5)$$

We can conclude that, to perform the same cooling operation, a $4306 \, \text{m}^2$ surface heat exchanger (!) and a $81026 \, \text{kg/h}$ air flow, i.e. more than 5 times the cooling tower air flow, are required.

This counterexample clearly highlights the benefits of a direct heat transfer, when possible, with respect to an indirect one. Moreover, the ΔT_{min} condition considerably limits the feasibility of the cooling operation performed with a heat exchanger.

3.3   Conclusions

This chapter, dealing with a classical cooling tower design problem, emphasizes all its potential compared to the classical heat exchange process. This operation is nevertheless constrained by the possibility to contact the two phases and by the thermodynamic equilibrium condition given by the wet bulb temperature.

In the industrial domain the need to solve the opposite problem could often occur as well. Given the already existing column and the air properties, the gas flowrate should be calculated for a desired outlet temperature. However, once mastered the design procedure, the opposite solution is straightforward.

In conclusion it can be stated that the cooling tower unit, even being not very different from a desorption column, is worth being studied as a standalone topic.

3.4   C++ Code

```cpp
#define BZZ_COMPILER 3
#include <BzzMath.hpp>

void Coolingtower(BzzVector &y, double z, BzzVector &dy);

FILE *res;

//DATA
double Lin = 20000;                 // kg/h
double Gdry = 15000;                // kg/h
double P = 760;                     // mmHg
double TinH2O = 50;                 // C
double ToutH2O = 25;                // C
double Tinair = 20;                 // C
double Z = 0.5;                     // relative humidity
double Cu = 0.26;                   // kcal/kg*K
double CpL = 1;                     // kcal/kg*K
double DHev = 580;                  // kcal/kg
double hga = 4000;                  // kcal/m/h/K
double hla = 30000;                 // kcal/m/h/K
double A = -49.705;                 // Pev 1st coeff mmHg vs C
double B = 2.71;                    // Pev 2nd coeff mmHg vs C

double Usair = 0.62*(A+B*Tinair)/P;
double Uair = Z*Usair;
double Kua = hga/Cu;
double L0 = 19292;                  // kg/h

void main(void)
{
double height = 0.;
double delta = 0.01;

res = fopen("Res.ris","w");
BzzVector y0(5,Uair,ToutH2O,Tinair,L0,Gdry);
printf("\nINITIAL CONDITIONS:\tU=%e\tTl=%e\tTg=%e\tL=%e\tG=%e",...
        ... y0[1],y0[2],y0[3],y0[4],y0[5]);
BzzVector y, dy;
BzzOdeNonStiff o(y0, height, Coolingtower);
y = y0;

for (int i = 0; i <= 1000; i++)
{
height = double(i);
printf("\nHEIGHT=%d\tU =%e\tTl =%e\tTg =%e\tL =%e\tG =%e", ...
        ... i,y[1],y[2],y[3],y[4],y[5]);
fprintf(res,"\n%e\t%e\t%e\t%e\t%e\t%e",height,y[1],y[2],y[3], ...
        ... y[4],y[5]);
o.SetInitialConditions(y, height);
y = o(height + delta);
}
fclose(res);
}

void Coolingtower(BzzVector &y, double z, BzzVector &dy)
{
// U=y[1], Tl=y[2], Tg=y[3], L=y[4], G=y[5]

double Ti=(hla*y[2]+hga*y[3]+Kua*DHev*(y[1]-0.62*A/P))/...
        ...(hla+hga+Kua*DHev*0.62*B/P);
double Ui=0.62*(A+B*Ti)/P;

dy[1] = Kua*(Ui - y[1]) / Gdry * delta;
```

```
dy[2] = hla*(y[2] - Ti) / y[4] / CpL * delta;
dy[3] = hga*(Ti - y[3]) / Gdry / Cu * delta;
dy[4] = Kua*(Ui - y[1]) * delta;
dy[5] = Kua*(Ui - y[1]) * delta;
}
```

Part II
Non-conventional Separation Units

Chapter 4
Multiple-Effect Evaporation

Abstract Multiple-effect evaporation is the cheapest and simplest unit operation aimed to either recover the solvent from a solution or concentrate the solution itself. Several system configurations exist and the optimal one from a Net Income point of view can be selected for each specific case. By mean of a desalination case study cocurrent and countercurrent schemes are compared and advantages and disadvantages of both of them are highlighted and commented from an operability and an economic perspective.

Keywords Concentration · Solvent recovery · Vaporization · Countercurrent · Economics

Evaporation and vaporization are two separation processes carried out using the same procedure but to achieve different goals. They're both performed by the equipment called "evaporator" where a liquid solution is fed. Vaporization has the purpose of recovering the vapor phase, i.e. the solvent (usually water), while evaporation is aimed at separating the solid phase or concentrating emulsions and suspensions.

Every single effect requires an external heat duty to vaporize part of the solvent that exits from the top of the evaporator; therefore the higher the solvent flowrate to be vaporized is, the higher the heat duty consumption is.

However, the solvent should be recovered in liquid phase, i.e. it needs to be condensated downstream the evaporator, wasting this way its latent heat and paying for an additional cooling duty.

The multiple-effect evaporator principle is based on splitting the solvent vaporization in several steps, recovering this way the otherwise wasted enthalpy of the vapor phase leaving each step to provide the heat duty required by the next one, considerably reducing the system energy consumption.

Thus, by adding more effects, a higher COP and a decrease of the operating costs can be achieved at the expenses of the investment costs.

Therefore if the desired product is the concentrated solution, whose revenue doesn't depend on N (number of effects), the net income shows a maximum, i.e. there exists an optimal N value that is usually about 2–3.

On the other hand if the desired product is the vaporized solvent the optimal number of effects is the highest possible, where the "possibility" constraint is given by the feasibility of the heat exchange between vapor and liquid phases in each unit and by the viscosity increase of the liquid solution due both to concentration increase and to temperature decrease.

4.1 Multiple-Effect Evaporator Design

A 10000 kg/h stream of acqueous solution with a non-volatile solute concentration of 5% w/w has to be concentrated in a simple cocurrent three effects evaporator. The solution inlet temperature is 90.0 °C and the LP steam temperature in the first effect is 140.0 °C.

Evaluate flowrates, temperatures and concentrations of the system for a 50% final solute concentration.

The physical properties and design data are reported in Table 4.1, the solution specific heat is supposed to be constant despite the concentration increase.

The boiling-point elevation follows the correlation here below:

$$\Delta T_{eb}(°C) = 60 \cdot x(w/w) \tag{4.1.1}$$

Solution

The system process scheme is shown in Fig. 4.1.

The missing degree of freedom of the system is satisfied by the final concentration specification. However, since several units in series need to be solved and data are

Table 4.1 Physical properties and design data

	Property	Value	Unit
	$c_P^{solution}$	1	$kcal/(kg \cdot K)$
	ΔH_{ev}^{steam}	511.5	kcal/kg
1st effect	A_1	150	m^2
	$\Delta H_{ev,1}^{vap}$	521.6	kcal/kg
	U_1	800	$kcal/(m^2 \cdot h \cdot K)$
2nd effect	A_2	150	m^2
	$\Delta H_{ev,2}^{vap}$	532.2	kcal/kg
	U_2	600	$kcal/(m^2 \cdot h \cdot K)$
3rd effect	A_3	150	m^2
	$\Delta H_{ev,3}^{vap}$	563.0	kcal/kg
	U_3	300	$kcal/(m^2 \cdot h \cdot K)$

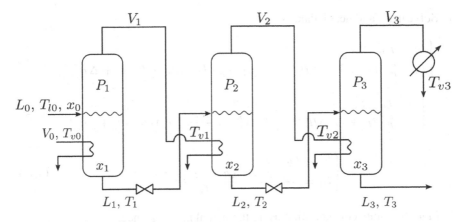

Fig. 4.1 Multiple-effect evaporator process scheme

available at the inlet while the specification is given at the outlet of the system, a trial-and-error approach is required.

The steam flowrate has been selected as manipulated variable since it doesn't depend on the others resulting this way the easiest to manage. Anyway every unknown variable could be chosen keeping unchanged the final results. Moreover, thanks to the Microsoft Excel "Goal seek" function, there's no need for several iterations, the correct solution can be quickly found.

Before going any further with the solution it can be noticed that the heat exchangers surface area of each effect are the same in order to ensure the interchangeability of them.

Furthermore it is worth remarking that the latent heat recovered by the vapors condensation is lower than the one required to vaporize the solvent since no solute is present anymore. On the other hand the condensation temperature is lower than the boiling one, therefore the latent heat value should be higher. Considering then it constant does not substantially bias the results.

The resolution methodology for a single effect will be explained in detail here below. It has to be repeated for each effect and the final x_3 value should be compared to the concentration specification in order to know whether reduce or increase the external steam flowrate.

1. First of all, given the heat duty, the temperature difference in the heat exchanger should be calculated:

$$V_0 \cdot \Delta H_{ev}^{steam} = U_1 \cdot A_1 \cdot \Delta T_1 \iff \Delta T_1 = \frac{V_0 \cdot \Delta H_{ev}^{steam}}{U_1 \cdot A_1} \qquad (4.1.2)$$

2. Calculate the boiling temperature of the solution:

$$T_1 = T_{v_0} - \Delta T_1 \qquad (4.1.3)$$

3. Solve mass and heat balances:

$$\begin{cases} L_0 = L_1 + V_1 \\ L_0 \cdot c_P^L \cdot T_{L_0} + V_0 \cdot \Delta H_{ev}^{steam} = L_1 \cdot c_P^L \cdot T_1 + V_1 \cdot (c_P^L \cdot T_1 + \Delta H_{ev,1}^{vap}) \end{cases} \Longleftrightarrow$$

$$\begin{cases} V_1 = L_0 - L_1 \\ L_0 \cdot c_P^L \cdot T_{L_0} + V_0 \cdot \Delta H_{ev}^{steam} = L_1 \cdot c_P^L \cdot T_1 + (L_0 - L_1) \cdot (c_P^L \cdot T_1 + \Delta H_{ev,1}^{vap}) \end{cases} \Longleftrightarrow$$

$$\begin{cases} V_1 = L_0 - L_1 \\ L_1 = \dfrac{L_0 \cdot c_P^L \cdot T_{L_0} + V_0 \cdot \Delta H_{ev}^{steam} - L_0 \cdot (c_P^L \cdot T_1 + \Delta H_{ev,1}^{vap})}{-\Delta H_{ev,1}^{vap}} \end{cases} \qquad (4.1.4)$$

4. Find the solute concentration using the partial mass balance:

$$L_0 \cdot x_0 = L_1 \cdot x_1 \Longleftrightarrow x_1 = \frac{L_0 \cdot x_0}{L_1} \qquad (4.1.5)$$

5. Calculate the boiling-point elevation using the given correlation:

$$\Delta T_{eb,1} = 60 \cdot x_1 \qquad (4.1.6)$$

6. The vapor condensation temperature results then to be:

$$T_{v_1} = T_1 - \Delta T_{eb,1} \qquad (4.1.7)$$

Once the single effect is completely solved, the same procedure needs to be repeated for the second and third ones.

The objective function can be defined as:

$$F_{obj} = (x_{3,spec} - x_{3,calc})^2 \qquad (4.1.8)$$

However, the mass fraction is a relatively small number that is quite sensitive to small variation compared to the Microsoft Excel "Goal seek" function tolerance. A more accurate result can be obtained using a flowrate based objective function.

Hence the outlet liquid flowrate can be calculated by imposing the partial mass balance on the whole system control volume:

$$L_0 \cdot x_0 = L_3 \cdot x_3 \Longleftrightarrow L_3 = \frac{L_0 \cdot x_0}{x_3} \qquad (4.1.9)$$

and a new objective function can be defined:

$$F'_{obj} = (L_{3,spec} - L_{3,calc})^2 \qquad (4.1.10)$$

Table 4.2 Vapor pressure correlation coefficients

C_1	$-5.8002206 \cdot 10^3$
C_2	1.3914993
C_3	$-4.8640239 \cdot 10^{-2}$
C_4	$4.1764768 \cdot 10^{-5}$
C_5	$-1.4452093 \cdot 10^{-8}$
C_6	6.5459673

For the sake of completeness the pressure of each effect has been calculate by using the following correlation[1]:

$$ln(P_{ev}(Pa)) = \frac{C_1}{T(K)} + C_2 + C_3 \cdot T + C_4 \cdot T^2 + C_5 \cdot T^3 + C_6 \cdot ln(T) \quad (4.1.11)$$

where the C_i constants are those reported in Table 4.2 and whose validity range is $[0, 200]\,°C$.

The detailed solution and system variables values are reported in Table 4.3.

4.2   Countercurrent Configuration Comparison

It would be interesting to compare the cocurrent solution to the countercurrent one (cf. Fig. 4.2). However, to estimate the countercurrent system variables the whole set of equations should be solved at once. This is due to the fact that even supposing a vapor flowrate value, there are not enough data to uniquely define each effect in series since vapor temperature would still be unknown.

To do this, a linear equation system solver is required. A simple C++ code has been implemented and the results are shown in Table 4.4.

As it can be noticed, in the countercurrent configuration the liquid solution goes through the effects in the opposite direction of the pressure gradient, that means pumps instead of less expensive valves should be employed. Moreover, the heat duty saved with respect to the cocurrent configuration is not particularly relevant (5.6% ca.).

Anyway, even if more expensive, the countercurrent configuration is generally preferred to the cocurrent one thanks to the possibility of directly managing the duty of the effect with the highest solute concentration. However, if thermolabile solution has to be treated, they would be damaged at high concentration by the high temperature of the last effect, thus the cocurrent configuration remains the only one possible.

---

[1] ASHRAE Handbook - Fundamentals 2002, Chap. 6.

Table 4.3 Multiple-effect evaporator results

	Variable	Value	Unit
Feed	L_0	10000	kg/h
	T_{L_0}	90.0	°C
	x_0	0.05	kg/kg
Duty	V_0	3540	kg/h
	T_{v_0}	140.0	°C
1st effect	ΔT_1	15.1	°C
	T_1	124.9	°C
	L_1	7198	kg/h
	V_1	2802	kg/h
	x_1	0.07	kg/kg
	$\Delta T_{eb,1}$	4.2	°C
	T_{v_1}	120.7	°C
	P_1	2.01	atm
2nd effect	ΔT_2	16.2	°C
	T_2	104.5	°C
	L_2	4175	kg/h
	V_2	3022	kg/h
	x_2	0.12	kg/kg
	$\Delta T_{eb,2}$	7.2	°C
	T_{v_2}	97.3	°C
	P_2	0.91	atm
3rd effect	ΔT_3	35.8	°C
	T_3	61.6	°C
	L_3	1000	kg/h
	V_3	3175	kg/h
	x_3	0.5	kg/kg
	$\Delta T_{eb,3}$	30.0	°C
	T_{v_3}	31.6	°C
	P_3	0.046	atm

4.3 Additional Remarks

An additional interesting case study could have concerned the identification of the optimal number of effects. In this problem the additional degree of freedom should be satisfied by mean of an economic optimization based on both capital and operating costs, i.e. equipment and external steam duty price, over the plant lifetime span. As already mentioned in the introduction, if the concentrated solution is the desired product, the net income function shows a maximum for a definite number of effect (cf. Fig. 4.3).

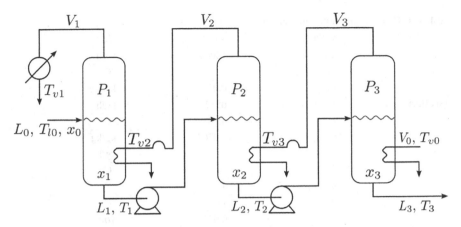

Fig. 4.2 Countercurrent multiple-effect evaporator process scheme

On the other hand, if the solvent needs to be recovered, the maximum net income is apparently obtained for an infinite number of effects (cf. Fig. 4.4). However, if we take into account that by increasing the number of effects:

- Temperature in the last effect decreases with a higher risk of crystallization and fouling over the heat exchanger tubes;
- Pressure in the last effect decreases causing the need for a thicker unit;
- Viscosity in the last effect increases because of the combined effect of lower temperature and higher concentration, affecting the pumping costs;

capital costs assessment can be corrected accordingly. After a certain number of effects, the actual net income function starts then decreasing showing also for this case a trend with a maximum as plotted in Fig. 4.5.

4.4 Conclusions

Multiple-effect evaporation is the most simple and widespread unit operation aimed either to concentrate a solution or to recover the solvent. In the former case, if the desired product is the pure solute, it could be used as a method to enhance crystallization.

The proposed case study consists in the design of a unit and shows the advantages of each of the two main configurations with respect to the operating conditions. Due to the constant trend of the driving force along the process, the countercurrent configuration results in a higher effectiveness as usually happens for separation operations. In both cases the final concentration specification fulfills the degree of freedom related to the external steam duty.

Table 4.4 Countercurrent multiple-effect evaporator results

	Variable	Value	Unit
Feed	L_0	10000	kg/h
	T_{L_0}	90	°C
	x_0	0.05	kg/kg
1st effect	L_1	6552	kg/h
	T_1	35.8	°C
	x_1	0.0763	kg/kg
	V_1	3448	kg/h
	T_{v_1}	31.2	°C
	$\Delta T_{eb,1}$	4.6	°C
	ΔT_1	11.7	°C
	P_1	0.045	atm
2nd effect	L_2	3924	kg/h
	T_2	55.0	°C
	x_2	0.1274	kg/kg
	V_2	2628	kg/h
	T_{v_2}	47.4	°C
	$\Delta T_{eb,2}$	7.7	°C
	ΔT_2	17.0	°C
	P_2	0.107	atm
3rd effect	L_3	1000	kg/h
	T_3	102.0	°C
	x_3	0.5	kg/kg
	V_3	2924	kg/h
	T_{v_3}	72.0	°C
	$\Delta T_{eb,3}$	30	°C
	ΔT_3	38.0	°C
	P_3	0.336	atm
Duty	V_0	3342	kg/h
	T_{v_0}	140	°C

Even if each operation requires an ad hoc design, this chapter outlines a complete overview on all the main aspect of multiple-effect evaporation to ensure a full understanding of the topic.

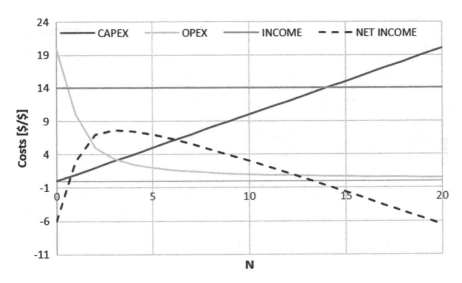

Fig. 4.3 Economic assessment—solution concentration

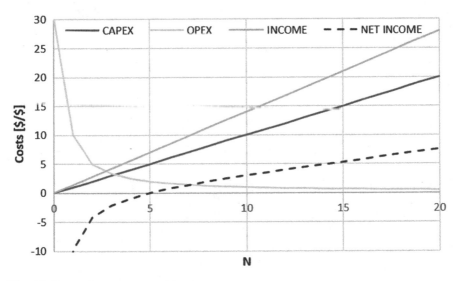

Fig. 4.4 Economic assessment—solvent recovery

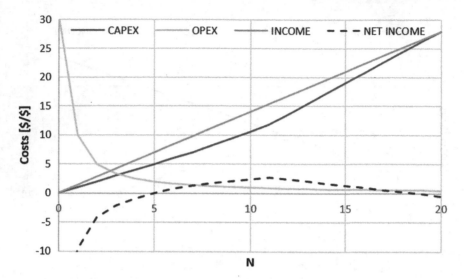

Fig. 4.5 Economic assessment—solvent recovery corrected

4.5   C++ Code

```
#define BZZ_COMPILER 3
#include <BzzMath.hpp>

// DATA
double L0 =10000;            // kg / hr
double T10 =90;              // C
double x0 =0.05;            // kg / kg
double Cpl =1;              // kcal / kg / C
double Tv0 =140;            // C
double dHev0 =511.5;        // kcal / kg
double dHev1 =563;          // kcal / kg
double dHev2 =532.2;        // kcal / kg
double dHev3 =521.6;        // kcal / kg
double U1 =800;             // kcal / m2 / hr / K
double U2 =600;             // kcal / m2 / hr / K
double U3 =300;             // kcal / m2 / hr / K
double A=150;               // m2
double Keb =60;             // C
double x3 =0.5;             // kg / kg


void Multipleeffect(BzzVector &y, BzzVector &f);

void main(void)
{
BzzPrint("\n\nAccess Functions");

BzzVector y0(15);
// L1 = y[1]; L2 = y[2]; L3 = y[3]
y0[1] = 7197.; y0[2] = 4175.; y0[3] = 1000.;

// V0 = y[4]; V1 = y[5]; V2 = y[6]; V3 = y[7];
y0[4] = 3540.; y0[5] = 2802.; y0[6] = 3022.; y0[7] = 3175.;

// T11 = y[8]; T12 = y[9]; T13 = y[10];
y0[8] = 125.; y0[9] = 105.; y0[10] = 61.;

// Tv1 = y[11]; Tv2 = y[12]; Tv3 = y[13];
y0[11] = 120.; y0[12] = 97.; y0[13] = 32.;

// x1 = y[14]; x2 = y[15];
y0[14] = 0.07; y0[15] = 0.12;

BzzNonLinearSystem nls(y0, Multipleeffect);
nls();
BzzPrint("\nIterations %d", nls.IterationCounter());
BzzPrint("\nMaximum Residual %e", nls.GetMaximumResidual());
BzzVector weights;
nls.GetWeights(&weights);
weights.BzzPrint("weights");
BzzPause();
BzzVector y, f;
nls.GetSolution(&y, &f);
y.BzzPrint("y solution");
f.BzzPrint("f solution");
BzzPause();
nls.BzzPrint("Results");
}



void Multipleeffect(BzzVector &y, BzzVector &f)
{
// Heat Exchanger Equation
```

```
f[1] = Tv0 - y[4]*dHev0 / U3 / A - y[10];
f[2] = y[13] - y[7]*dHev3 / U2 / A - y[9];
f[3] = y[12] - y[6]*dHev2 / U1 / A - y[8];

// Boiling Point Elevation
f[4] = y[11] + Keb*y[14] - y[8];
f[5] = y[12] + Keb*y[15] - y[9];
f[6] = y[13] + Keb*x3 - y[10];

// Overall Mass Balance
f[7] = L0 - y[1] - y[5];
f[8] = y[1] - y[2] - y[6];
f[9] = y[2] - y[3] - y[7];

// Heat Balance
f[10] = L0*Cpl*T10 + y[6]*dHev2 - y[1]*Cpl*y[8] - y[5]*(Cpl*y[8]+dHev1);
f[11] = y[1]*Cpl*y[8] + y[7]*dHev3 - y[2]*Cpl*y[9] - y[6]*(Cpl*y[9]+dHev2);
f[12] = y[2]*Cpl*y[9] + y[4]*dHev0 - y[3]*Cpl*y[10] - y[7]*(Cpl*y[10]+dHev3);

// Component Mass Balance
f[13] = L0*x0 - y[1]*y[14];
f[14] = L0*x0 - y[2]*y[15];
f[15] = L0*x0 - y[3]*x3;
}
```

Chapter 5
Filtration

Abstract Filtration is the less expensive and most established process intended to mechanically separate the suspended solids from a slurry. Among the several existing filtration technologies the most suitable one can be detected according to the feed to be treated. It is usually operated as a batch process, thus an optimal residence time can be found in order to optimize the daily productivity. Whether a higher driving force is required the centrifugal configuration can be selected and modeled after performing an accurate operating conditions check.

Keywords Pressure-driven · Batch · Scheduling · Centrifugal

Filtration is the most common and, in the majority of cases, the cheapest unit operation aimed to separate a fluid phase from solid particles. It is often employed upstream more refined operations in order to pretreat the stream by removing the bigger impurities. The separation occurs because the so called "filter" mechanically intercepts the solid phase letting the fluid one pass.

Different filtration typologies, that is different applications, exist according to the solid particles size to be separated (cf. Table 5.1). Other classifying criteria are the driving force, the filtration mechanism, the purpose or batch vs continuous operation.

The choice of the most suitable material for the filter depends on the nature of the process stream and of the filter itself.

The most common filtration models are the so called "constant pressure drop" models. They assume a non-compressible cake, whose properties don't change in time, and they are valid for non-compressible precipitates and constant pressure filtration. On the other hand the geometrical properties of the panel filter change. Among this models the Darcy one is the most popular by far; it attributes the decrease of filtration velocity to the increase of the height of the panel due to the accumulation of solids causing a higher resistance to flow. On the other hand the cross-sectional area available for filtration remains equal to the filter area.

Table 5.1 Filtration processes classification

Size [μm]	Mixture type	Filtration typology	Example of particles
>100	Suspension	Macrofiltration	Solid particles
0.1/100	Suspension	Microfiltration	Bacteria, algae and fungi
0.001/0.1	Solution	Ultrafiltration	Virus, colloids
0.0001/0.001	Solution	Reverse osmosis	Big molecules

5.1 Filtration Time Optimization

A homogeneous suspension undergoes a batch filtration cycle with a $\Delta P = 3\,\text{kg/cm}^2$, forming in 1 h a 20 mm cake and producing 6 m^3 of filtrate.

After each cycle the cake is washed with water; the washing phase occurs with the same operating conditions consuming 1 m^3 of water.

Before each filtration and washing operation, 2 min are required to load the filter; after each of them the cake is drained for 3 min. Then filter assembling operations globally require 6 min.

It is assumed the filtrate and washing water to have the same properties and the filtering leaf resistance to be negligible. The hypotheses of Darcy equation applies.

Requests are namely:

• Calculate the daily filtrate volume;
• Calculate the daily filtrate volume whether a 12 mm cake was formed corresponding to a 3.6 m^3 of filtrate volume. The washing water versus filtrate volume ratio as well as all the other operating conditions stay unchanged;
• Optimize the filtration time.

Solution I

A simplified model of cake filtration is shown in Fig. 5.1.

The filtration model is based on the Poiseuille equation:

$$Q = \frac{\Delta P \cdot A}{32 \cdot \mu \cdot \frac{L}{D^2}}$$

(5.1.1)

where

• Q is the liquid flowrate;
• ΔP is the pressure difference across the filter;
• A is the effective cross sectional area, i.e. $\varepsilon \cdot A_{panel}$;
• μ is the viscosity of the fluid;

Fig. 5.1 Cake filtration
scheme

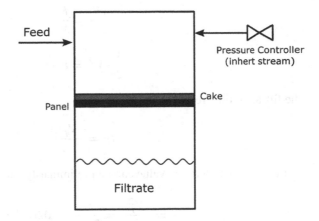

- L is the length of the channels, i.e. cake + panel;
- D is the holes diameter.

The Darcy model is based on the hypothesis that the only parameter that changes during the filtration operation is $L = L_0 + \Delta L(t)$. This length increase is due to the solid deposition on the filter and it can be better defined by mean of the material balance:

$$\Delta L = \frac{V_{cake}}{A} = \frac{m_{cake}}{\rho_{cake} \cdot A} = \frac{C_s \cdot V}{\rho_{cake} \cdot A} = K \cdot V \qquad (5.1.2)$$

The Poiseuille equation then becomes:

$$Q = \frac{dV}{dt} = \frac{\Delta P \cdot A}{32 \cdot \mu \cdot \frac{(L_0 + K \cdot V)}{D^2}} \qquad (5.1.3)$$

Its integration is straightforward:

$$(L_0 + K \cdot V) \cdot dV = \frac{\Delta P \cdot A}{32 \cdot \frac{\mu}{D^2}} \cdot dt \iff L_0 \cdot V + \frac{1}{2} \cdot K \cdot V^2 = \frac{\Delta P \cdot A}{32 \cdot \frac{\mu}{D^2}} \cdot t$$
$$(5.1.4)$$

For a long filtration time the linear term is negligible and the equations can be rewritten as:

$$\frac{dV}{dt} = \frac{\Delta P \cdot A}{32 \cdot \frac{\mu}{D^2} \cdot K \cdot V} \qquad (5.1.5)$$

$$\frac{1}{2} \cdot K \cdot V^2 = \frac{\Delta P \cdot A}{32 \cdot \frac{\mu}{D^2}} \cdot t \qquad (5.1.6)$$

It is worth noticing that, since all the terms but V are constant during the operation, the K constant could include as many of them as needed. The shortest way to write the Darcy model is indeed:

$$\frac{dV}{dt} = \frac{K}{V} \tag{5.1.7}$$

$$\frac{1}{2} \cdot V^2 = K \cdot t \tag{5.1.8}$$

The filtration time results to be then:

$$t_F = \frac{V_F^2}{2 \cdot K} \tag{5.1.9}$$

Using this equation the K value can be preliminarily calculated:

$$K = \frac{V_F^2}{2 \cdot t_F} = \frac{6^2}{2 \cdot 60} = 0.3 \, \text{m}^6/\text{min} \tag{5.1.10}$$

This value is strictly related to the operating conditions that are the same for every problem request. The washing phase is characterized by a washing velocity equal to the final filtration velocity:

$$\frac{dV_W}{dt} = \frac{dV_F}{dt}\Big|_{t_F} = \frac{K}{V_F} \tag{5.1.11}$$

The washing time is given by simple integration:

$$V_W = \frac{K}{V_F} \cdot t_W \iff \frac{V_W}{V_F} = \frac{K}{V_F^2} \cdot t_W = \frac{t_W}{2 \cdot t_F} \iff \frac{V_F}{V_W} = \alpha = \frac{2 \cdot t_F}{t_W} \tag{5.1.12}$$

The calculation of t_W is then straightforward:

$$t_W = \frac{2 \cdot t_F \cdot V_W}{V_F} = \frac{2 \cdot t_F}{\alpha} = \frac{2 \cdot 60}{6} = 20 \, \text{min} \tag{5.1.13}$$

Each filtration cycle is scheduled as shown in Table 5.2. Each filtration+washing cycle implies 16 min of downtime. The cycle duration is obtained by summing up all the terms:

$$t_{cycle} = t_{down} + t_F + t_W = 16 + 60 + 20 = 96 \, \text{min} \tag{5.1.14}$$

Table 5.2 Filtration cycle [min]

Assembling	Load	Filtration	Draining	Water load	Washing	Draining
6	2	t_F	3	2	t_W	3

Assuming the filtration process to occur during the whole day, the number of cycles per day can be evaluated as:

$$n = \frac{hours/day \cdot minutes/hours}{t_{cycle}} = \frac{24 \cdot 60}{96} = 15 \, cycles/day \qquad (5.1.15)$$

Finally the daily productivity results to be:

$$P\,[m^3/day] = V_F \cdot n = 6 \cdot 15 = 90 \, m^3/day \qquad (5.1.16)$$

Solution II

The second case study is similar to the first one with a different filtration phase duration. Filtrate conditions are the same and we can check it by the proportionality between cake formation and filtrate volume

$$C_s \propto \frac{\Delta L}{V_F} = \frac{20\,mm}{6\,m^3} = \frac{12\,mm}{3.6\,m^3} \qquad (5.1.17)$$

Therefore the K value stays unchanged; α as well remains the same according to the hypotheses. The filtration time is then given by:

$$t_F = \frac{V_F^2}{2 \cdot K} = \frac{3.6^2}{2 \cdot 0.3} = 21.6 \, min \qquad (5.1.18)$$

On the other hand the washing time is:

$$t_W = \frac{2 \cdot t_F}{\alpha} = 7.2 \, min \qquad (5.1.19)$$

Thus the cycle duration results to be $t_{cycle} = 44.8$ min. It corresponds to about 32 cycles per day, that is $P = 115.7 \, m^3/day$.

Solution III

The results previously obtained show that the filtration duration affects the daily productivity. A longer t_F reduces the number of daily cycles increasing the amount of filtrate per cycle. On the other hand a small t_F substantially increases the number of cycles per day while the contribution of t_{down} on the total time becomes more significant. Thus it is reasonable to think that the P versus t_F function shows a trend with a maximum. It is then worth looking for the filtration optimal duration.

In order to optimize the productivity with respect to the filtration time the second variable in the first function should be made explicit as follows:

$$t_{cycle} = t_{down} + t_F + t_W = t_{down} + t_F + \frac{2}{\alpha} \cdot t_F \qquad (5.1.20)$$

The daily number of cycles results then:

$$n = \frac{24 \cdot 60}{t_{down} + (1 + \frac{2}{\alpha}) \cdot t_F} \qquad (5.1.21)$$

The filtrate volume is given by:

$$V_F = \sqrt{2 \cdot K \cdot t_F} \qquad (5.1.22)$$

Finally the productivity is obtained as the product of the two:

$$P = V_F \cdot n = \sqrt{2 \cdot K \cdot t_F} \cdot \frac{24 \cdot 60}{t_{down} + (1 + \frac{2}{\alpha}) \cdot t_F} \qquad (5.1.23)$$

The sufficient and necessary condition of productivity maximization corresponds to setting its derivative with respect to t_F equal to 0:

$$\frac{dP}{dt_F} = \frac{24 \cdot 60 \cdot \sqrt{2 \cdot K}}{(t_{down} + (1 + \frac{2}{\alpha}) \cdot t_F)^2} \cdot \left(\frac{1}{2 \cdot \sqrt{t_F}} \cdot (t_{down} + \left(1 + \frac{2}{\alpha}\right) \cdot t_F) - \sqrt{t_F} \cdot \left(1 + \frac{2}{\alpha}\right) \right) = 0$$
$$(5.1.24)$$

The first part is always positive, therefore the equation can be simplified as:

$$\frac{1}{2 \cdot \sqrt{t_F}} \cdot \left(t_{down} + \left(1 + \frac{2}{\alpha}\right) \cdot t_F \right) - \sqrt{t_F} \cdot \left(1 + \frac{2}{\alpha}\right) = 0 \Longleftrightarrow \qquad (5.1.25)$$

$$\Longleftrightarrow \left(t_{down} + \left(1 + \frac{2}{\alpha}\right) \cdot t_F \right) - 2 \cdot t_F \cdot \left(1 + \frac{2}{\alpha}\right) = 0 \Longleftrightarrow \qquad (5.1.26)$$

$$\Longleftrightarrow t_{down} - t_F \cdot \left(1 + \frac{2}{\alpha}\right) = 0 \Longleftrightarrow t_F^{opt} = \frac{t_{down}}{(1 + \frac{2}{\alpha})} \qquad (5.1.27)$$

Thus, for this case study, the optimal filtration time is $t_F^{opt} = 12 \, \text{min}$ and the corresponding washing time is $t_W = 4 \, \text{min}$. The overall cycle duration is then 32 min, that is 45 cycles per day, and the filtrate volume is $V_F = 2.68 \, \text{m}^3$ corresponding to a daily productivity $P = 120.74 \, \text{m}^3/\text{day}$.

For the sake of completeness the trend of P as a function of t_F is shown in Fig. 5.2.

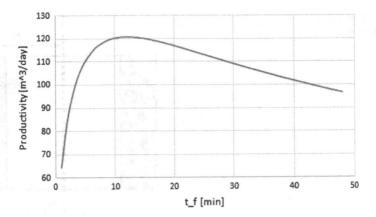

Fig. 5.2 Productivity versus filtration time

5.2 Centrifugal Filtration

Given a basket loaded with a liquid to be filtered according to the data reported in Table 5.3 and the scheme in Fig. 5.3, calculate:

- The decantation time under the hypothesis of solid particles laminar flow;
- The time according to the Darcy law required to filter the 98% of the liquid.

Solution I

Centrifugal filtration is analougous to the gravitational one with the main difference that, in the former, the fluid phase takes advantage of the centrifugal force to overcome

Table 5.3 Physical properties and operating conditions

Property	Symbol	Value	Unit
Solid concentration	C_s	100	kg/m³
Loaded fraction	ξ	0.5	m³/m³
Particle diameter	D_P	$1 \cdot 10^{-3}$	m
Channel diameter	D	$5 \cdot 10^{-5}$	m
Solid density	ρ_s	2200	kg/m³
Water density	ρ_w	1000	kg/m³
Angular velocity	ω	1	round/s
Viscosity	μ	10^{-3}	Pa · s
Basket height	H	0.4	m
Radius	R	0.25	m
Cake void fraction	ε	0.4	1

Fig. 5.3 Centrifugal
filtration scheme

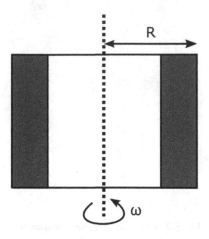

the pressure drops across the cake. While in gravitational filtration the driving force can be improved by pumping inherts in the system, in centrifugal filtration the same effect can be obtained by speeding up the rotation.

The liquid to be filtered takes a cylindrical shape parallel to the basket walls due to the high rotational speed, therefore the lateral surface corresponds to the filtration area.

This operation is characterized by two phenomena:

- Decantation: solid particles velocity is higher than the liquid one. Thus, the first phenomenon is the deposition of the solid cake on the filter surface. After that, the already clarified liquid passes across the maximum possible resistance from the very beginning;
- Filtration: the liquid phase carries with it the solid particles towards the filter because of the driving force. There exists proportionality between the filtrate volume and the cake deposition.

If $t_D \ll t_F$, as usually happens for centrifugal driving force, decantation occurs.

In order to know the prevailing phenomenon each characteristic time should be calculated.

When decantation occurs, after a short transient, the particle moves with a constant velocity. Under this condition two forces acting on the particle, namely the centrifugal force and the viscosity, are equal to each other. This balance can be expressed as:

$$\frac{1}{2} \cdot f \cdot \rho_w \cdot v^2 \cdot \frac{\pi \cdot D_P^2}{4} = (\rho_s - \rho_w) \cdot \frac{\pi \cdot D_P^3}{6} \cdot \omega^2 \cdot r \qquad (5.2.1)$$

being:

$$f = \frac{24}{Re} = \frac{24 \cdot \mu}{\rho_w \cdot v \cdot D_P} \qquad (5.2.2)$$

The Stokes law is then obtained:

$$\frac{1}{2} \cdot \frac{24 \cdot \mu}{\cancel{\rho_w} \cdot \cancel{v} \cdot \cancel{D_R}} \cdot \cancel{\rho_w} \cdot \cancel{v^2} \frac{\cancel{\pi} \cdot D_P^{\cancel{2}}}{4} = (\rho_s - \rho_w) \cdot \frac{\cancel{\pi} \cdot D_P^3}{6} \cdot \omega^2 \cdot r \qquad (5.2.3)$$

$$v = \frac{dr}{dt} = \frac{(\rho_s - \rho_w) \cdot D_P^2 \cdot \omega^2 \cdot r}{18 \cdot \mu} = K \cdot r \qquad (5.2.4)$$

By integrating the velocity equation between the inner and outer radius the decantation time can be estimated:

$$\frac{dr}{r} = K \cdot dt \iff t_D = \frac{1}{K} \cdot ln\left(\frac{R}{R_i}\right) \qquad (5.2.5)$$

In order to solve the equation the parameters K and R_i need to be evaluated. Given the constant density of the solution, the inner radius is calculated by imposing the volume balance:

$$\xi \cdot V_{basket} = V_{loaded} \iff \xi \cdot \cancel{\pi} \cdot R^2 \cdot \cancel{H} = \cancel{\pi} \cdot R^2 \cdot \cancel{H} - \cancel{\pi} \cdot R_i^2 \cdot \cancel{H} \iff R_i^2 = (1 - \xi) \cdot R^2 \qquad (5.2.6)$$

that finally is:

$$R_i = \sqrt{(1 - \xi)} \cdot R = 0.1768\,m \qquad (5.2.7)$$

On the other hand the decantation constant K is given by:

$$K = \frac{(\rho_s - \rho_w) \cdot D_P^2 \cdot \omega^2}{18 \cdot \mu} = 2.63\,s^{-1} \qquad (5.2.8)$$

so finally:

$$t_D = \frac{1}{2.63} \cdot ln(\frac{0.25}{0.1768}) = 0.132\,s \qquad (5.2.9)$$

For the sake of completeness the cake thickness can be assessed. The solid quantity is equal to:

$$m_{cake} = C_s \cdot V_{loaded} = C_s \cdot (1 - \xi) \cdot \pi \cdot R^2 \cdot H = 3.926\,kg \qquad (5.2.10)$$

$$V_{cake} = \frac{m_{cake}}{\rho_s} = \frac{3.926}{2200} = 1.784 \cdot 10^{-3}\,m^3 \qquad (5.2.11)$$

Thus the cake thickness is obtained approximating:

$$\Delta L = \frac{V_{cake}}{S_l} = \frac{V_{cake}}{2 \cdot \pi \cdot R \cdot H} = 2.84\,mm \qquad (5.2.12)$$

Solution II

The Darcy model is exactly the same explained in Sect. 5.1 for gravitational filtration. However, the accurate selection of the equation form to use requires particular care. Neglecting L_0, the most general one is:

$$\frac{dV}{dt} = \frac{\Delta P \cdot A}{32 \cdot \frac{\mu}{D^2} \cdot \frac{C_s \cdot V}{\rho_{cake} \cdot A}} \tag{5.2.13}$$

The only terms that change during the operation are ΔP and V. The cake density is given by:

$$\rho_{cake} = \rho_s \cdot (1 - \varepsilon) + \rho_w \cdot \varepsilon = 1720 \, kg/m^3 \tag{5.2.14}$$

Therefore the Darcy constant can be defined as:

$$K_D = 32 \cdot \frac{\mu}{D^2} \cdot \frac{C_s}{\rho_{cake} \cdot A^2} = 1.89 \cdot 10^6 \, Pa \cdot s/m^6 \tag{5.2.15}$$

The pressure drops due to the centrifugal field correspond to:

$$\Delta P = \frac{1}{2} \cdot \rho_w \cdot \omega^2 \cdot (r_0^2 - r^2) = \frac{1}{2} \cdot \rho_w \cdot \omega^2 \cdot \frac{(V_0 - V)}{\pi \cdot H} = K_P \cdot (V_0 - V) \tag{5.2.16}$$

where:

$$K_P = \frac{1}{2} \cdot \frac{\rho_w \cdot \omega^2}{\pi \cdot H} = 15708 \, Pa/m^3 \tag{5.2.17}$$

$$V_0 = V_{loaded} = 0.03926 \, m^3 \tag{5.2.18}$$

Finally, the Darcy equation states as:

$$\frac{dV}{dt} = \frac{K_P}{K_D} \cdot \frac{V_0 - V}{V} = K \cdot \frac{V_0 - V}{V} \tag{5.2.19}$$

with $K = 0.0083 \, m^3/s$. In order to evaluate the filtration time the equation needs to be integrated. The following variable change could be useful:

$$\frac{V}{V_0} = a \Longleftrightarrow dV = V_0 \cdot da \tag{5.2.20}$$

Then the differential equation becomes:

$$V_0 \cdot da = K \cdot \frac{1-a}{a} \cdot dt \Longleftrightarrow dt = \frac{V_0}{K} \cdot (\frac{1}{1-a} - 1) \cdot da \tag{5.2.21}$$

integrating:

$$t_F = \frac{V_0}{K} \cdot (ln(\frac{1}{1-a}) - a)|_0^{0.98} = 13.82\,\mathrm{s} \tag{5.2.22}$$

In conclusion, since $t_D \ll t_F$, our expectation about the system operation under decantation conditions is confirmed.

5.3 Conclusions

The case study proposed in this chapter has the purpose to explain the Darcy model for pressure-driven filtration and to discuss the operation scheduling optimization. Scheduling, usually related to maintenance operations, is a typical feature of batch processes and we proved that an optimal filtration time for daily productivity maximization exists for the filtration process.

The second example refers to the centrifugal configuration: this operation takes advantage of the density difference between liquid and solid phases. Differently from the centrifugal sedimentation, it cannot be performed really in continuous due to the impossibility to continuously remove the cake. For these reasons this operation is more suitable for refined filtration processes involving relatively clarified liquids and smaller particles diameter.

Chapter 6
Drying

Abstract Drying is the most general expression for a set of operations aimed to remove the moisture contained in solid matrix products by mean of heat transfer. Due to the several possible configurations a wide range of models exists; they share in common heat and mass balances both in case of direct and indirect heat transfer. Internal diffusion can be either taken into account or neglected and a proper discretization should be used to model this phenomenon.

Keywords Moisture removal · Direct heat transfer · Discretization

Drying is a unit operation aimed to reduce the moisture content in a solid phase by evaporation/phase transition. The goal is achieved providing heat to the moist solid in two possible ways:

- Direct heat transfer: a hot gas (usually dry air) provides heat to the solid and strips away the vapor;
- Indirect heat transfer:
  - Conduction: the contact with a hot surface provides the required heat;
  - Radiation: a high temperature source generates the required heat;
  - Electric arc: an electric current passing through the solid heats it up according to the Joule effect.

Analogously to other processes dealing with evaporation, both evaporating and phase transition conditions affect the operation.

The selected drier technology depends on the properties of the material to be treated; drying in general is performed on materials that are not subject to degradation or property loss at high temperature.

Table 6.1 Physical properties and operating conditions

Property	Symbol	Value	Unit
Initial temperature	T_0	20	°C
Initial moisture	W_0	0.4	kg/kg
Slice thickness	H	0.02	m
Slice side	L	0.1	m
Bread density	ρ_s	400	kg/m^3
Bread heat capacity	c_P	3200	J/(kg · K)
Bread thermal conductivity	k	0.8	W/(m · K)
Heat transfer coefficient	h	116.278	W/(m^2 · K)
Water vaporization enthalpy	ΔH_{ev}	2176.72 ·10^3	J/kg
Operating conditions			
Pressure	P	101325	Pa
Gas temperature	T_g	180	°C

6.1 Bread Drying Example

A square slice of bread, whose properties are listed in Table 6.1, is dried using hot air under the operating conditions reported in the same Table.

The following assumptions are valid:

- Only convective heat transfer through the upper surface is considered;
- Phase transition conditions;
- No diffusion limitations for the vapor transfer;
- Solid physical properties remain constant during the operation.

The requests are as follows:

- Write down the material and heat balances needed to describe the system both under heating and phase transition conditions. For this purpose a discretization of the total thickness into three equal and homogeneous control volumes is suggested;
- Use the obtained model to evaluate the time required to completely dry the bread slice. Plot and discuss the water content trend both locally and globally.

The Euler method is suggested for the numerical integration of the drying model differential equations with a maximum $\Delta t = 10$ s.

Solution I

First of all a few remarks are worth to be made about the Euler method. The Euler method is a first-order numerical procedure for solving ordinary differential equations with a given initial value. It is the most basic explicit method for numerical integration of ODEs. Given the Cauchy problem:

$$y'(t) = f(t, y(t)) \qquad y(t_0) = y_0 \tag{6.1.1}$$

The integration step is defined as:

$$y_{n+1} = y_n + h \cdot f(t_n, y_n) \tag{6.1.2}$$

It is based on the derivative approximation to the value that the function assumes in the previous point of the discretized integration interval.

The bread slice is then divided into three homogeneous "subslices" and the drying occurs only under phase transition conditions once the upper subslice is completely dried. Without this discretization the whole slice can be considered as homogeneous during the drying operation that in this case would be an oversimplifying hypothesis. However, in other case studies it could result in a realistic assumption as well as a much denser discretization could be required in order to overlap the real phenomenon according to the internal transport resistances.

To sum up, the occurring physical phenomena are in the order:

1. Heating of the upper subslice;
2. Drying of the upper subslice;
3. Heating of the middle subslice;
4. Drying of the middle subslice;
5. Heating of the lower subslice;
6. Drying of the lower subslice;

Therefore model equations will be written referring to each subslice one by one (cf. Fig. 6.1).

1. Upper subslice

During heating phase no mass transfer occurs while heat is provided by the gas and conducted through the bread:

$$\begin{cases} \frac{\partial m_w}{\partial t} = 0 \\ m_s \cdot c_P \cdot \frac{\partial T_1}{\partial t} = h \cdot A \cdot (T_g - T_1) - \frac{k}{H_s} \cdot A \cdot (T_1 - T_2) \end{cases} \tag{6.1.3}$$

On the other hand when $T_1 = 100\,°\text{C}$ evaporation occurs:

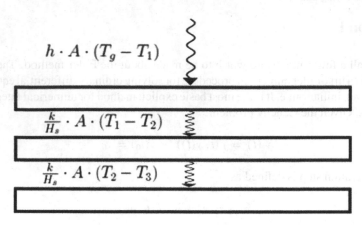

$$h \cdot A \cdot (T_g - T_1)$$

$$\frac{k}{H_s} \cdot A \cdot (T_1 - T_2)$$

$$\frac{k}{H_s} \cdot A \cdot (T_2 - T_3)$$

Fig. 6.1 Heat transfer through the slice

$$\begin{cases} \frac{\partial m_w}{\partial t} \cdot \Delta H_{ev} = -h \cdot A \cdot (T_g - T_1) + \frac{k}{H_s} \cdot A \cdot (T_1 - T_2) \\ T_1 = 100\,^{\circ}\mathrm{C} \end{cases} \qquad (6.1.4)$$

Once the upper slice is completely dried, the "heating phase" equations are applied again.

2. Middle subslice

During the heating phase no mass transfer occurs while heat is provided by conduction from the upper slice and conducted to the lower one:

$$\begin{cases} \frac{\partial m_w}{\partial t} = 0 \\ m_s \cdot c_P \cdot \frac{\partial T_2}{\partial t} = \frac{k}{H_s} \cdot A \cdot (T_1 - T_2) - \frac{k}{H_s} \cdot A \cdot (T_2 - T_3) \end{cases} \qquad (6.1.5)$$

When $T_2 = 100\,^{\circ}\mathrm{C}$ model equations become:

$$\begin{cases} \frac{\partial m_w}{\partial t} \cdot \Delta H_{ev} = \frac{k}{H_s} \cdot A \cdot (T_1 - T_2) - \frac{k}{H_s} \cdot A \cdot (T_2 - T_3) \\ T_2 = 100\,^{\circ}\mathrm{C} \end{cases} \qquad (6.1.6)$$

Once the middle slice is completely dried, the "heating phase" equations are applied again.

3. Lower subslice

During the heating phase no mass transfer occurs while heat is provided by conduction from the middle slice:

$$\begin{cases} \frac{\partial m_w}{\partial t} = 0 \\ m_s \cdot c_P \cdot \frac{\partial T_3}{\partial t} = \frac{k}{H_s} \cdot A \cdot (T_2 - T_3) \end{cases} \qquad (6.1.7)$$

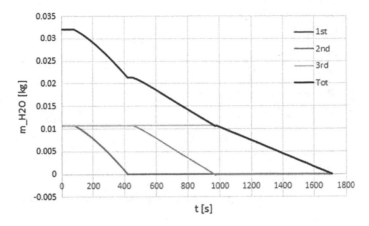

Fig. 6.2 Moisture content

Then when $T_3 = 100\,°C$ model equations become:

$$\begin{cases} \frac{\partial m_w}{\partial t} \cdot \Delta H_{ev} = \frac{k}{H_s} \cdot A \cdot (T_2 - T_3) \\ T_3 = 100\,°C \end{cases} \tag{6.1.8}$$

Once the lower slice is completely dried as well, the drying process is completed.

Solution II

Once all the model equations have been defined, they can be numerical integrated by using a Microsoft Excel Worksheet to implement the Euler method. The results thus obtained are shown in Figs. 6.2 and 6.3.

The complete drying is attained after about 29 min.

It's worth noticing that the heating phases are relatively short and they become shorter during the operation since the lower slices keep on being heated even during the drying phase of the upper ones.

By going from the top to the bottom of the slice the water content decreases more slowly, this phenomenon is due to the fact that the dried part thickness increases providing a higher heat transfer resistance. The first layer indeed requires about 7 min to completely dry, the second layer about 9 min and the last layer about 13 min.

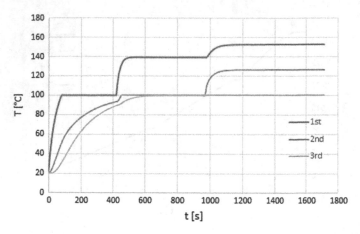

Fig. 6.3 Temperature profiles

6.2 Conclusions

Several different configurations aimed to the same separation are included under the name of "drying processes". In this chapter the simplest and the most general case of a solid object drying in a hot dry gas stream is explained and numerically solved in detail.

In this example the object to be dried, i.e. the slice of bread, has been divided into three sub-slices. This numerical assumption reflects the heterogeneous conditions along the slice thickness and the delay during the heating phase. In case of a lower thermal diffusivity, a denser discretization would be required to provide a more realistic model. On the other hand in case of negligible internal resistances, no discretization would be needed and the total moisture content would show a uniform linear trend during time.

Chapter 7
Spray Drying

Abstract When suspensions, aerosols or slurries contain solid particles that cannot be mechanically dried a spray dryer should be employed. Widely spread in food and pharma industry, spray dryers are suitable to ensure low residence time at high temperature. The unit height is designed according to the residual moisture specification by mean of an ODEs system arising from mass and heat balances.

Keywords Slurry · Powder · Atomization · Food · Drying chamber

Spray dryers are continous direct convective heat transfer dryers. They are used to remove a fluid phase (usually water) from suspensions, aerosols or slurries containing solid particles that cannot be mechanically dried, that are heat-sensitive and cannot be exposed to high temperatures for a long time or even fluid phases containing ultrafine particles that could melt and agglomerate if dried under different conditions.

Spray dryers are indeed suitable for food and pharma products due to the short residence time in the hot zone of the unit and due to the protection from very high temperatures ensured by the liquid film over the particles. This liquid film causes the steady state operating temperature to be equal to the wet bulb temperature of the drying air. Moreover, spray dryers are suitable for color pigments whose size needs to be the same as the original solid particles one or for slurries with very fine dispersed particles and non-newtonian behaviour even with very low moisture content, such as clays.

Spray dryers consist of a large cylindrical or conical chamber in which the material to be processed is atomized and sprayed by mean of nozzles. Liquid drops containing the solid particles are mixed with a gas flow hot enough to provide the required heat to let the whole liquid phase evaporate. After this operation, the gas is cooled down and separated from solid particles; the air-solid separation is partially obtained at the bottom of the chamber thanks to gravity. For smaller particles, cyclones or more effective processes are used, such as sleeves filters or electrostatic separator.

Drying chambers are typically big empty equipment: they should be high enough to ensure the contacting time required to obtain the desired residual moisture; the diameter needs to be large enough to avoid contact between particles and wall, with resulting precipitation and decantation.

7.1 Spray Dryer Design for Powdered Milk Production

Powdered milk should be produced by mean of a spray dryer whose nozzle atomizes 0.2 mm diameter drops with an initial velocity $v_P^0 = 0.3$ m/s.

Milk and air physical properties are listed in Table 7.1.

Given an equipment diameter equal to 5.5 m, the calculation of the equipment height for a final moisture equal to 0.005 kg/kg is requested.

Solution

The general scheme of a spray dryer is shown in Fig. 7.1. The operation involves two phases, namely the solid particle phase and the dry air one; water is transferred from the solid matrix to the gas stream by evaporation, its latent heat is compensated by the convective heat flux provided by the hot air stream. Mass and heat balances for the single particle can be easily written as:

$$\frac{dm_P}{dt} = -K_P \cdot S_P \cdot (P^0(T_P) - P_w) \tag{7.1.1}$$

$$m_P \cdot c_P^m \cdot \frac{dT_g}{dt} = h \cdot S_P \cdot (T_g - T_P) - K_P \cdot S_P \cdot (P^0(T_P) - P_w) \cdot \Delta H_{ev} \tag{7.1.2}$$

On the other hand, when dealing with the gas phase, it should be taken into account that the gas stream passing through the slice dz exchanges vapor and heat with all the particles contained in the volume $A \cdot dz$. Given the parameter η, defined as the number of particles per unit volume, the number of particles per unit time is given by:

Table 7.1 Milk and air properties

Milk	Value	Unit	Air	Value	Unit
Q_{milk}	1 750	kg/h	G_{dry}	72 000	kg/h
T_P^0	303	K	T_g^0	403	K
fat	4.76	% (kg/kg)	P	1	atm
ρ_P	1000	kg/m^3	μ	$2.3 \cdot 10^{-5}$	kg/(m · s)
c_P^m	1	kcal/(kg · K)	c_P^g	0.25	kcal/(kg · K)
ΔH_{ev}	540	kcal/kg	D_{va}	$1.8 \cdot 10^{-5}$	m^2/s
Antoine		mmHg vs K	k	$8 \cdot 10^{-6}$	kcal/(m · s · K)
A	18.3036				
B	3816.44				
C	−46.13				

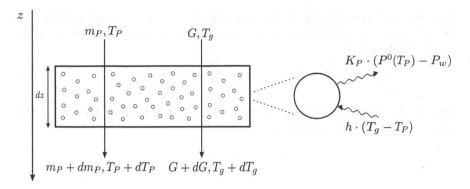

Fig. 7.1 Spray dryer model

$$\frac{\eta \cdot dV}{dt} = \eta \cdot \frac{A \cdot dz}{dt} = \eta \cdot A \cdot v_P \tag{7.1.3}$$

Therefore mass and heat balances for the gas phase can be written as:

$$\frac{dG}{dt} = K_P \cdot S_P \cdot (P^0(T_P) - P_w) \cdot \eta \cdot A \cdot v_P \tag{7.1.4}$$

$$G \cdot c_P^g \cdot \frac{dT_P}{dt} = -h \cdot S_P \cdot (T_g - T_P) \cdot \eta \cdot A \cdot v_P \tag{7.1.5}$$

The momentum balance is the only model equation missing. Gas phase velocity can be considered constant since neither the increase in volume nor friction on the wall are relevant. On the contrary particle velocity considerably changes before achieving the steady state value, thus momentum balance on the particle could be included or not, according to the duration of the transient with respect to the total residence time in the spray dryer. In case of relevant transient, given the relative velocity $v_s = v_P - v_g$, the momentum balance, states as:

$$\frac{d(m_P \cdot v_s)}{dt} = \rho_P \cdot V_P \cdot g - \rho_g \cdot V_P \cdot g - \frac{1}{2} \cdot f \cdot \rho_g \cdot v_s \cdot |v_s| \cdot \pi \cdot \frac{D_P^2}{4} \tag{7.1.6}$$

For a falling sphere in a viscous fluid under laminar flow regime $f = \frac{24}{Re}$, that is:

$$\frac{dm_P}{dt} \cdot v_s + m_P \cdot \frac{dv_s}{dt} = (\rho_P - \rho_g) \cdot V_P \cdot g - \frac{1}{2} \cdot \frac{24 \cdot \mu}{\rho_g \cdot |v_s| \cdot D_P} \cdot \rho_g \cdot v_s \cdot |v_s| \cdot \pi \cdot \frac{D_P^2}{4} \tag{7.1.7}$$

thus, simplifying:

$$- K_P \cdot S_P \cdot (P^0(T_P) - P_w) \cdot v_s + m_P \cdot \frac{dv_s}{dt} = (\rho_P - \rho_g) \cdot V_P \cdot g - 3 \cdot \mu \cdot v_s \cdot \pi \cdot D_P \tag{7.1.8}$$

Sometimes the contribution due to the change in mass is neglected and the relative velocity, for a negligible transient, is considered constant and calculated as the asymptotical one:

$$v_s^\infty = \frac{(\rho_P - \rho_g) \cdot g \cdot D_P^2}{18 \cdot \mu} = 0.947 \,\text{m/s} \tag{7.1.9}$$

Finally, since the design of the spray dryer is related to the equipment height, a sixth differential equation is required:

$$\frac{dz}{dt} = v_P = v_s + v_g \tag{7.1.10}$$

The spray dryer results then to be modeled by a set of six differential equations, that means six initial conditions are required. The first one is the initial particle mass that is a function of the initial particle diameter, i.e. of the atomizer performance. It can be evaluated as:

$$m_P^0 = \rho_P \cdot V_P^0 = \rho_P \cdot \frac{\pi \cdot (D_P^0)^3}{6} = 4.19 \cdot 10^{-9} \,\text{kg} \tag{7.1.11}$$

The inlet particle temperature T_P^0 is given as well as the inlet dry air flowrate and temperature G_{dry} and T_g^0. The initial relative velocity should be assessed as the difference between the initial particle velocity and the gas one; the value of the former is given while the latter one, considered constant, is calculated as:

$$v_g = \frac{Q_{air}^V}{A} = \frac{Q_{air}^W}{\rho_g} \cdot \frac{4}{\pi \cdot D^2} = 0.96 \,\text{m/s} \tag{7.1.12}$$

where the gas density is obtained by using the ideal gas EoS:

$$\rho_g = \frac{P}{R \cdot T_g^0} \cdot MW_{air} = 0.877 \,\text{kg/m}^3 \tag{7.1.13}$$

then we obtain:

$$v_s^0 = v_P^0 - v_g = -0.66 \,\text{m/s} \tag{7.1.14}$$

The last initial condition for the axial coordinate could be simply set as $z^0 = 0 \,\text{m}$.

Now that the values to start the ODEs integration are known, it should be assessed when to stop. The specification is indeed given as final moisture of the particle but no moisture appears in the equations that means we have to correlate one of the differential variables to the specification. The water content is clearly related to the mass of the particle; given the fat fraction of the milk, the initial moisture of each particle can be assessed:

$$W_{in} = \frac{Q_l}{Q_{fat}} = \frac{Q_{milk} - Q_{fat}}{Q_{fat}} = \frac{Q_{milk} \cdot (1 - fat)}{Q_{milk} \cdot fat} = \frac{(1 - fat)}{fat} = 20 \, \text{kg/kg}$$

$$(7.1.15)$$

and then the mass of the dry particle is obtained as:

$$m_P = m_{dry} + m_{water} = m_{dry} \cdot (1 + W) \Longleftrightarrow m_{dry} = \frac{m_P^0}{(1 + W_{in})} = 1.99 \cdot 10^{-10} \, \text{kg}$$

$$(7.1.16)$$

By using the same equation for the final moisture specification, the particle final mass can be evaluated:

$$m_P^{fin} = m_{dry} \cdot (1 + W_{fin}) = 2 \cdot 10^{-10} \, \text{kg} \qquad (7.1.17)$$

The very last missing step is the evaluation of transport coefficients. In order to do that, the required dimensionless numbers need to be calculated:

$$Re = \frac{\rho_g \cdot v_s \cdot D_P}{\mu} \qquad (7.1.18)$$

$$Pr = \frac{\mu \cdot c_P^g}{k} \qquad (7.1.19)$$

$$Sc = \frac{\mu}{\rho_g \cdot D_{va}} \qquad (7.1.20)$$

For a spherical geometry the Nusselt and Sherwood numbers can be obtained as:

$$Nu = 2 + 0.4 \cdot Re^{0.5} \cdot Pr^{1/3} \qquad (7.1.21)$$

$$Sh = 2 + 0.4 \cdot Re^{0.5} \cdot Sc^{1/3} \qquad (7.1.22)$$

Therefore the mass and heat transfer coefficients result from the correlations:

$$h = \frac{Nu \cdot k}{D_P} \, [\text{kcal}/(\text{m}^2 \cdot \text{s} \cdot \text{K})] \qquad (7.1.23)$$

$$K_C = \frac{Sh \cdot D_{va}}{D_P} \, [\text{m/s}] \qquad (7.1.24)$$

However, the mass transfer coefficient here above is on a concentration basis while mass vs. pressure basis is required. It can be converted by mean of the following equivalences:

$$K_C^w \cdot (C^0 - C_w) = K_C \cdot MW_w \cdot \left(\frac{P^0}{R \cdot T_g} - \frac{P_w}{R \cdot T_g} \right) \Longleftrightarrow K_P^w = \frac{K_C}{R \cdot T_g} \cdot MW_w$$

$$(7.1.25)$$

Finally the system of ordinary differential equations can be solved by writing a simple code in C++ and stop the integration for $m_P = 2 \cdot 10^{-10}$, that is $t = 5.79\,s$ and $z = 10.91\,m$. Velocity and temperature profiles are shown in Figs. 7.2a, b while the particle mass and the axial coordinate trends are shown in Figs. 7.3a, b. Even if Fig. 7.2a is the one related to velocity profile, Fig. 7.3b results more immediate to determine whether the transient is relevant or not. The design decision is related to the height of the unit and the transient describes the delay after which velocity can be considered constant, that means the longer the transient the biggest the difference between simplified and accurate model results. Although velocity is stable after about half a second, its value is considerably lower than the asymptotical one only for the first 0.2 s, after which the axial coordinate trend is almost linear. This means that in this case the simplified model would have provided almost the same result as the detailed one.

7.2 Conclusions

Spray drying is a common process seldom discussed in detail. As shown in this application, its simple model equations are strictly related to the fluid dynamic conditions. The cocurrent case study has been analyzed in deep, while the countercurrent configuration can be easily obtained by reversing the gas velocity vector.

This model has been derived by assuming a single falling drop, which means that no interactions between the particle and other particles, as the chamber wall, have been taken into account. This interactions become even more relevant with a countercurrent gas flow, but such a detailed model is rarely required for common applications at industrial level.

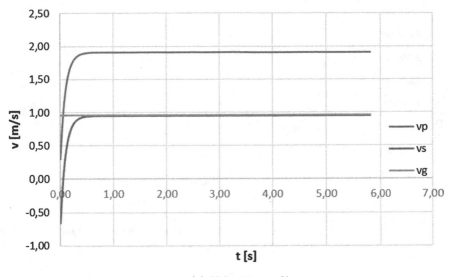

(a) Velocity profiles

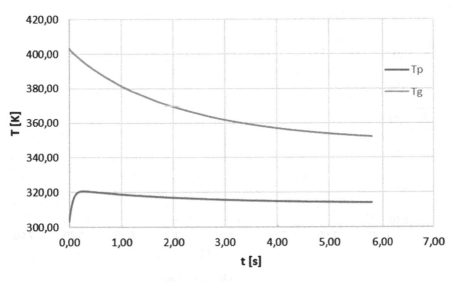

(b) Temperature profiles

Fig. 7.2 Velocity and temperature profiles

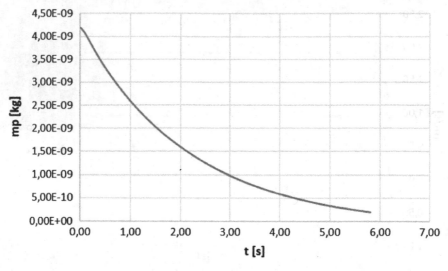

(a) Particle mass

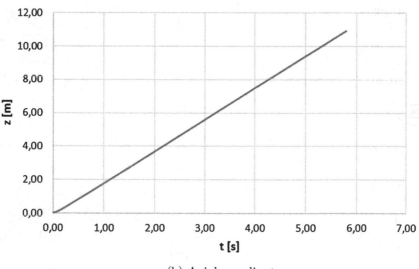

(b) Axial coordinate

Fig. 7.3 Specification and axial coordinate trends

7.3   C++ Code

```cpp
#define BZZ_COMPILER 3
#include "BzzMath.hpp"
#include "cmath"

void Spraydryer(BzzVector &y, double t, BzzVector &dy);

FILE *res;

// Physical properties
double Tp0 = 303;              // K
double Tg0 = 403;              // K
double P = 1;                  // atm
double CpL = 1;               // kcal/kg/K
double CpG = 0.25;            // kcal/kg/K
double DHev = 540;            // kcal/kg
double rhoL = 1000;          // kg/m^3
double rhoG = P/0.0821/Tg0*29;      // kg/m^3
double muG = 2.3*1.0e-5;      // kg/m/s
double kG = 8.*1.0e-6;        // kcal/m/s/K
double Diff = 1.8*1.e-5;      // m^2/s

// Antoine equation
double A = 18.3036;
double B = 3816.44;
double C = -46.13;

// Process stream
double Qmilk = 1750. / 3600;        // kg/h -> kg/s
double fat = 4.76 / 100;       // kg/kg
double Qfat = Qmilk*fat;       // kg/s
double Ql = Qmilk - Qfat;      // kg/s
double Win = Ql / Qfat;        // kg/kg
double Wout = 0.005;           // kg/kg
double dp = 2 * 1.0e-4;        // m
double vp0 = 0.3;          // m/s
double mp0 = rhoL*3.14/6*pow(dp,3); // kg
double mdry = mp0 / (1 + Win);      // kg
double mout = mdry*(1 + Wout);      // kg
double nAvp = Qmilk / mp0;     // drops/s


// Gas stream
double Gdry = 20;             // kg/s
double D = 5.5;              // m

// Momentum Balance
double vg = Gdry/rhoG/3.14/BzzPow2(D)*4;    // m/s
double vs = (rhoL-rhoG)*BzzPow2(dp)*9.81/18/muG;// m/s
double vp = vg + vs;           // m/s
double vs0 = vp0 - vg;         // m/s

// Mass and heat transfer
double Re = rhoG*vs*dp / muG;           // Reynolds
double Pr = muG*CpG / kG;           // Prandtl
double Sc = muG / rhoG / Diff;          // Schmidt
double Nu = 2.+0.4*pow(Re,0.5)*pow(Pr,0.333);   // Nusselt
double Sh = 2.+0.4*pow(Re,0.5)*pow(Sc,0.333);   // Sherwood
double h = Nu*kG / dp;           // kcal/m^2/s/K
double Kc = Sh*Diff / dp;       // m/s
double Kpw = Kc/0.0821/Tg0*18;           // kg/m^2/s/atm

void main(void)
{
```

```
double time = 0.;
double delta = 1.e-3;
res = fopen("Res.ris", "w");
BzzVector y0(6, mp0, 0, Tp0, Tg0, vs0, 0);

printf("\nINITIAL CONDITIONS:\tmp=%e\tGvap=%e\tTp=%e\tTg=%e\tvp=%e\tz=%e",...
        ... y0[1],y0[2],y0[3],y0[4],y0[5],y0[6]);

BzzVector y, dy;
BzzOdeNonStiff o(y0, time, Spraydryer);
y = y0;

for (int i = 0; i <= 800; i++)
{
time = double(i);
printf("\nTIME=%d\tmp=%e\tGvap=%e\tTp=%e\tTg=%e\tvp=%e\tz=%e\tz=%e",...
        ... i, y[1], y[2], y[3], y[4], y[5], y[6]);
fprintf(res, "\n%e\t%e\t%e\t%e\t%e\t%e\t%e",...
        ... time, y[1], y[2], y[3], y[4], y[5], y[6]);
o.SetInitialConditions(y, time);
y = o(time + delta);
}
fclose(res);
}

void Spraydryer(BzzVector &y, double t, BzzVector &dy)
{
// mp=y[1], Gvap=y[2], Tp=y[3], Tg=y[4], vp=y[5], z=y[6]

double Pw = 29/18*y[2]*P/Gdry;
double P0 = exp(A-B/(y[3]+C))/760;   // mmHg->atm
double Sp = 3.14*pow(pow((y[1]/rhoL)*6/3.14,0.3333),2);

dy[1] = Kpw*Sp*(Pw-P0)*delta ;
dy[2] = -Kpw*Sp*(Pw-P0)*nAvp*delta;
dy[3] = (h*Sp*(y[4]-y[3])+Kpw*Sp*(Pw-P0)*DHev)/y[1]/CpL*delta;
dy[4] = -h*Sp*(y[4]-y[3])*nAvp/(y[2]+Gdry)/CpG*delta;
dy[5] = ((1-rhoG/rhoL)*9.81-3*y[5]*muG*3.14*dp/mp0)*delta;
dy[6] = (y[5]+vg)*delta;
}
```

Chapter 8
Lyophilization

Abstract Due to the thermolability of food and pharmaceutical products the drying process cannot be performed at high temperature; lyophilization replaces then standard drying units taking advantage of sublimation instead of evaporation. Being freeze drying a batch operation, a suitable model is used to assess the operating residence time for given operating conditions. The critical importance of an optimal control system for the productivity maximization will be discussed by mean of an common industrial case study.

Keywords Sublimation · Freezing · Control · Irradiation

Freeze drying (or lyophilization) is a unit operation meant to remove moisture from a solid phase. It is a process analogous to the drying operation but for thermolabile solids, where evaporation is replaced by sublimation. For this reason operating conditions are below the water triple point (4.58 mmHg and 0.01 °C) and, as for drying, the water mass transfer occurs because of the fulfillment of two different conditions:

- Phase transition: sublimation depends only on the heat provided to the system and not on the external conditions, pressure should be properly controlled in order to keep constant the phase transition temperature;
- Pressure gradient: water partial pressure in the gas phase is lower than the equilibrium pressure, therefore mass transfer occurs even if the temperature of the system is lower than the sublimation one. The mass flow can be expressed as $J = K_P \cdot S \cdot (P_w - P^0(T))$.

Being sublimation an endothermic process, the solid surface temperature would decrease during time reducing the phase transition rate. Hence, heat should be constantly provided in order to keep the sublimation front temperature as high as possible (according to the T_s^{max} constraint).

Lyophilization is a batch process lasting even more than 30 h, therefore, even if simple, it results to be quite expensive.

The process cycle steps can be resumed as follows:

1. Product freezing;
2. Primary lyophilization by sublimation;

Table 8.1 Operating conditions and chemico-physical properties

Parameter	Symbol	Value	Unit
Plate temperature	T_P	120	°C
Sublimation temperature	T_F	−20	°C
Apple bed height	H	0.03	m
Moisture	W	0.9	kg/kg
Radiant heat transfer coefficient	σ	$4.87 \cdot 10^{-8}$	kcal/(m² · h · K⁴)
Property	Symbol	Value	Unit
Solid density	ρ_s	1000	kg/m³
Sublimation enthalpy	ΔH_s	680	kcal/kg
Solid conductivity	k	0.1	kcal/(m · h · K)
Maximum solid temperature	T_S^{max}	50	°C

3. Secondary lyophilization by desorption;
4. Reset of pressure and temperature conditions.

8.1 Apples Lyophilization

In food industry lyophilization is a widespread process and freeze dried fruit is a very common product worth to be studied in deep.

This case study refers to apples lyophilization under the operating conditions listed in Table 8.1.

The requests are namely:

- Calculate the required time to completely lyophilize the apples;
- Find the plate temperature that gives a dry solid surface temperature no greater than 50 °C;
- Propose a temperature program for the plate (starting from $T_P = 120$ °C) respecting the aforementioned condition (i.e. $T_S < 50$ °C).

Solution I

The sublimation front model that will be used to solve this lyophilization problem is based on the existence of a surface sliding across the solid and separating the dried part from the wet one. It relies on two assumptions:

- Steady state of the dried solid phase;
- Constant geometrical properties: the cross section A and height H of the solid do not change during the process.

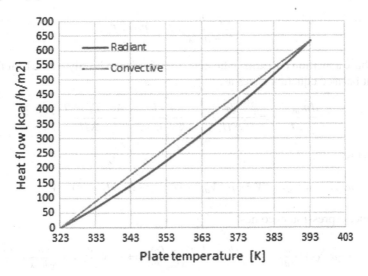

Fig. 8.1 Radiant versus convective heat flux

First of all we can formally linearize the radiant heat flux in a convective-like one in the range of interest [50, 120] °C (i.e. [323, 393] K) (cf. Fig. 8.1).

$$J_I = \sigma \cdot A \cdot (T_P^4 - T_S^4) = h \cdot A \cdot (T_P - T_S) \qquad (8.1.1)$$

Thus a convective heat transfer coefficient can be defined as:

$$h = \frac{\sigma \cdot (T_P^4 - T_S^4)}{T_P - T_S} = 9.03\,\text{kcal}/(\text{m}^2 \cdot \text{h} \cdot \text{K}) \qquad (8.1.2)$$

On the other hand the heat flux due to conduction between the upper surface and the front is given by:

$$J_C = \frac{k}{x} \cdot A \cdot (T_S - T_F) \qquad (8.1.3)$$

where x is the sublimation front height.

Due to the steady state hypothesis, J_I and J_C have to be equal, hence an overall heat transfer coefficient can be defined as:

$$\frac{1}{U} = \frac{1}{h} + \frac{x}{k} \Longleftrightarrow U = \frac{k \cdot h}{k + h \cdot x} \qquad (8.1.4)$$

this coefficient depends on the front position and decreases with the progression of the process. The total heat flux results to be:

$$J_T = \frac{k \cdot h}{k + h \cdot x} \cdot A \cdot (T_P - T_F) \tag{8.1.5}$$

In phase transition conditions the mass transfer depends only on the heat flux and the heat balance equation states as:

$$\frac{dm_{water}}{dt} = -\frac{J_T}{\Delta H_s} = -\frac{k \cdot h}{k + h \cdot x} \cdot \frac{A \cdot (T_P - T_F)}{\Delta H_s} \tag{8.1.6}$$

Then being:

$$m_{water} = m_{dry} \cdot W = \rho \cdot V_{dry} \cdot W = \rho \cdot A \cdot (H - x) \cdot W \tag{8.1.7}$$

the former expression becomes:

$$\frac{dm_{water}}{dt} = -\rho \cdot A \cdot W \cdot \frac{dx}{dt} = -\frac{k \cdot h}{k + h \cdot x} \cdot \frac{A \cdot (T_P - T_F)}{\Delta H_s} \Longleftrightarrow \rho \cdot W \cdot \frac{dx}{dt} = \frac{k \cdot h}{k + h \cdot x} \cdot \frac{(T_P - T_F)}{\Delta H_s}$$
$$\tag{8.1.8}$$

Finally the differential equation should be integrated by separation of variables:

$$\int_0^x (k + h \cdot x) \cdot dx = \int_0^t \frac{k \cdot h \cdot (T_P - T_F)}{\rho \cdot W \cdot \Delta H_s} \cdot dt \Longleftrightarrow k \cdot x + \frac{1}{2} \cdot h \cdot x^2 = \frac{k \cdot h \cdot (T_P - T_F)}{\rho \cdot W \cdot \Delta H_s} \cdot t$$
$$\tag{8.1.9}$$

The equation obtained by integration relates the operation time to the front position. Therefore the answer to the first request can be obtained by imposing the complete lyophilization condition $x = H = 0.03 \, m$ corresponding to a residence time $t = 34.2 \, h$.

Solution II

The first step to perform it in order to solve the next request of the exercise is the calculation of the solid surface temperature by mean of the continuity of the heat fluxes, i.e. solid steady state hypothesis, as follows:

$$\frac{k}{x} \cdot A \cdot (T_S - T_F) = h \cdot A \cdot (T_P - T_S) \Longleftrightarrow T_S = \frac{k \cdot T_F + h \cdot x \cdot T_P}{k + h \cdot x} \tag{8.1.10}$$

The surface temperature results to be a weighted average temperature between the sublimation and the plate one.

In Fig. 8.2 the surface temperature under the given operating condition $T_P = 120\,°C$ has been plotted for the whole duration of the lyophilization process.

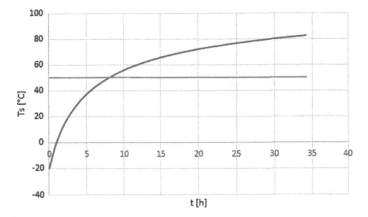

Fig. 8.2 Surface temperature for $T_P = 120\,°C$

It results evident that, by setting a plate temperature of $T_P = 120\,°C$, the maximum allowed surface temperature value for the integrity of the product is overcome after 8 h.

In order to avoid product degradation the plate temperature should be reduced. In doing so, the price of the product integrity is paid with a longer residence time. Being the system still the same, the previous model equations can be used and the T_P corresponding to a $T_S^{max} = 50\,°C$ can be easily determined by mean of the Microsoft Excel "Goal seek" function.

The result is $T_P = 75.83\,°C$ and the corresponding surface temperature trend can be observed in Fig. 8.3, the residence time increases this way from 34 to 50 h.

Solution III

The very last request of the problem leads to a compromise between the first two. We want the product not to degradate and, at the same time, we want to mitigate the residence time increase for the operation. The main idea is then to lower down the plate temperature when the surface temperature achieves the maximum allowed value.

A step T_P profile starting from $T_P = 120\,°C$ can be tested. From the Solution I paragraph it results that T_S^{max} is achieved after $t_1 = 8\,h$ for $x_1 = 0.011\,m$.

The plate temperature can be then reduced to 95 °C for instance. Due to the hypothesis of negligible transient, as soon as the plate temperature is reduced, the surface temperature is instantaneously reduced as well. Now there are two possible options:

1. The reference system can be shifted to the conditions x_1, t_1 and the previous equations can still be used;

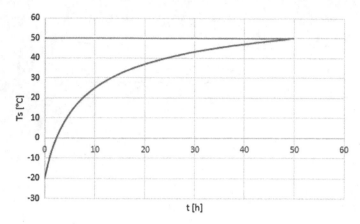

Fig. 8.3 Surface temperature for $T_P = 75.8\,^\circ\text{C}$

2. The integration of the differential equation can be performed in the new interval as follows:

$$\int_{x_1}^{x_2} (k + h \cdot x) \cdot dx = \int_{t_1}^{t_2} \frac{k \cdot h \cdot (T_P - T_F)}{\rho \cdot W \cdot \Delta H_s} \cdot dt \Longleftrightarrow \qquad (8.1.11)$$

$$\Longleftrightarrow k \cdot (x_2 - x_1) + \frac{1}{2} \cdot h \cdot (x_2^2 - x_1^2) = \frac{k \cdot h \cdot (T_P - T_F)}{\rho \cdot W \cdot \Delta H_s} \cdot (t_2 - t_1) \quad (8.1.12)$$

In this case the T_S^{max} is achieved at $t_2 = 16.3\,\text{h}$ and $x_2 = 0.017\,\text{m}$. Therefore the plate temperature needs to be stepped down again. If we want it to be the last step we should look for a T_P corresponding to $T_S = 50\,^\circ\text{C}$ for $x_3 = 0.3\,\text{m}$, but this problem has already been solved in Solution II and such a temperature is $T_P = 75.8\,^\circ\text{C}$. The balance can be now solved again finding the residence time $t_3 = 44.6\,\text{h}$.

In Table 8.2 all the results referring to the three considered operating conditions are listed and the plot corresponding to the last case is shown in Fig. 8.4.

It is worth remarking that, by properly manipulating the plate temperature, more than 10 % of the time can be saved respecting the degradation temperature constraint.

8.2 Optimal Plate Temperature Trend

The step T_P function described in the Sect. 8.1 third paragraph is an approximation of the best possible path for the plate temperature variable. Thanks to this example it can be deduced that the best operating conditions for the lyophilizer are those allowing the maximum surface temperature during the whole operation. Actually

Table 8.2 Lyophilization results for $T_P = 120\,°C$, $T_P = 75.8\,°C$ and variable T_P,

x (m)	t (h)	T_s (°C)	x (m)	t (h)	T_s (° C)	x (m)	t (h)	T_P (°C)	T_s (°C)
0	0	−20	0	0	−20	0	0	120	−20
0.0015	0.77	−3.29	0.0015	1.13	−8.56	0.0015	0.77	120	−3.29
0.003	1.65	9.85	0.003	2.41	0.43	0.003	1.65	120	9.85
0.0045	2.62	20.47	0.0045	3.83	7.70	0.0045	2.62	120	20.47
0.006	3.69	29.21	0.006	5.39	13.69	0.006	3.69	120	29.21
0.0075	4.86	36.55	0.0075	7.10	18.71	0.0075	4.86	120	36.55
0.009	6.13	42.79	0.009	8.95	22.97	0.009	6.13	120	42.78
0.0105	7.49	48.16	0.0105	10.94	26.65	0.0105	7.49	120	48.15
0.012	8.95	52.83	0.012	13.08	29.85	0.01107	8.03	120	50
0.0135	10.52	56.93	0.0135	15.36	32.66	0.01107	8.03	95	37.5
0.015	12.18	60.56	0.015	17.79	35.14	0.012	9.15	95	39.82
0.0165	13.93	63.79	0.0165	20.36	37.35	0.0135	11.06	95	43.19
0.018	15.79	66.69	0.018	23.07	39.34	0.015	13.08	95	46.17
0.0195	17.75	69.31	0.0195	25.93	41.13	0.0165	15.22	95	48.83
0.021	19.80	71.68	0.021	28.93	42.75	0.0172	16.28	95	50
0.0225	21.95	73.84	0.0225	32.07	44.23	0.0172	16.28	75.83	38.33
0.024	24.20	75.81	0.024	35.36	45.58	0.018	17.72	75.83	39.34
0.0255	26.55	77.63	0.0255	38.79	46.82	0.0195	20.57	75.83	41.13
0.027	29.00	79.30	0.027	42.37	47.96	0.021	23.57	75.83	42.75
0.0285	31.54	80.84	0.0285	46.08	49.02	0.0225	26.72	75.83	44.23
0.03	34.19	82.27	0.03	49.95	50	0.024	30.00	75.83	45.58
						0.0255	33.44	75.83	46.82
						0.027	37.01	75.83	47.96
						0.0285	40.73	75.83	49.02
						0.03	44.59	75.83	50

the plate temperature can be directly calculated given the surface one for every value of the front position x, i.e. of the residence time t, by inverting the Eq. 8.1.10:

$$\frac{k}{x} \cdot A \cdot (T_S - T_F) = h \cdot A \cdot (T_P - T_S) \Longleftrightarrow T_P = T_S + \frac{k \cdot (T_S - T_F)}{h \cdot x} \qquad (8.2.1)$$

This way, setting $T_S = 50\,°C$ from the very first moment, T_P shows a hyperbolic optimal trend as plotted in Fig. 8.5.

Actually the plate temperature cannot assume every desired value, being it constrained by technology limitations and energy consumption (i.e. profitability). However, these operating conditions allow for the estimation of the minimum theorical time required for the lyophilization process, that is about 39.5 h. This means that

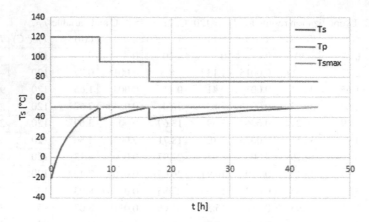

Fig. 8.4 Surface temperature for variable T_P

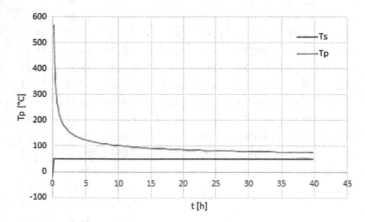

Fig. 8.5 Theorical optimal T_P path

even the best lyophilizer designed by the best engineer cannot freeze-dry the apples of this case study without degradating them with a lower residence time .

Thus, in order to work in real conditions, let's say that the maximum attainable plate temperature T_P is 120 °C. It can be then set to 120 °C until a lower value is required by the degradation constraint; the resulting plate and solid surface temperature paths are shown in Fig. 8.6 and the corresponding residence time is about 42.5 h. Moreover, the related C++ code is reported in the following paragraph. It can be finally concluded that this simple example shows how a good process control allows to save up to the 15% of the residence time for a batch process, that is a substantial increase in the process productivity.

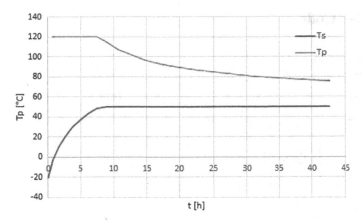

Fig. 8.6 Real optimal T_P path

8.3 Conclusions

A detailed analysis of the freeze drying process has been proposed through a series
of progressively more accurate case studies. Process optimization and control have
been discussed in deep and the results obtained for lyophilization can be generalized
to all batch processes.

Variables trends are the consequence of the steady state hypothesis, but if energy
accumulation in the dried phase is not negligible, a dynamic model is required and
a delay could be observed between an external perturbation and the front response.

8.4 C++ Code

```cpp
#define BZZ_COMPILER 3
#include <BzzMath.hpp>

void Lyophilization(BzzVector &y, double t, BzzVector &f);

FILE *res;

// DATA
double Tp = 120 + 273.15;                    //K
double Tf = -20 + 273.15;                    //K
double Ts = 50 + 273.15;                     //K
double Tsmax = 50 + 273.15;                  //K
double W = 0.9;                              //kg/kg
double H = 0.03;                             //m
double rho = 1000.;                          //kg/m3
double dHsub = 680.;                         //kcal/kg
double k = 0.1 / 60;                         //kcal/(m*min*K)
double sigma = 4.87e-8 / 60;                 //kcal/(m2*min*K^4)
double h = sigma*(BzzPow4(Tp)-BzzPow4(Ts))/(Tp-Ts);//kcal/(m2*min*K)

void main(void)
{
res = fopen("Res.ris", "w");
BzzVector y0(1, 0.), y;
BzzOdeNonStiff o(y0, 0., Lyophilization);
y = y0;

for (int i = 1; i <= 3000; i++)
{
o.SetInitialConditions(y, double(i));
y = o(double(i + 1));

Ts = (k*Tf + h*y[1] * Tp) / (k + h*y[1]);

Tp = Tsmax + k*(Tsmax - Tf) / h / y[1];
if (Tp > 120 + 273.15)
{
    Tp = 120 + 273.15;
}

printf("\nTIME =%d\tx =%e\t Ts =%e\t Tp =%e", i, y[1], Ts, Tp);
fprintf(res,"\n%d\t%e\t%e\t%e", i, y[1], Ts-273.15, Tp-273.15);

}
fclose(res);
}

void Lyophilization(BzzVector &y, double t, BzzVector &f)
{
    f[1] = ((Tp - Tf)*k*h) / (dHsub*rho*W) / (k + h*y[1]);
}
```

Chapter 9
Gas-Solid Separation

Abstract Gas streams can be purified from the suspended particles by mean of units with no moving parts and then no variables that can be manipulated: abatement chambers and cyclones. For a given separation efficiency the latter unit results to be much more compact than the former one and the accurate sizing procedure is discussed through a dedicated case study.

Keywords Abatement · Particulate · Centrifugal · Drift velocity · Particle diameter

The cyclone is a gas-solid separation unit with no moving parts. It takes advantage of the change in trajectory of the gas stream, tangentially fed, to separate the solid particles that keep moving in the tangential direction until their impact on the wall and their consequential fall.

The cyclone is a compact equipment that can achieve a 99% separation efficiency even with respect to small particles. However, for a given flowrate, the minimum particle size it is able to separate depends on its geometrical properties, causing the employ of two equal cyclones in series to be useless.

9.1 Cyclone Separation Effectiveness Calculation

A $0.25 \, m^3/s$ air stream with suspended particles is sent to a cyclone separator as the one shown in Fig. 9.1. Given the grains distribution reported in Table 9.1, the cyclone geometrical properties and the air stream physical ones listed in Table 9.2, the separation efficiency calculation is requested.

Solution

The cyclone separator behaviour is modeled as an equivalent abatement chamber.
First of all the linear velocity and distance are evaluated:

© The Author(s), under exclusive license to Springer Nature Switzerland AG 2020
A. Di Pretoro and F. Manenti, *Non-conventional Unit Operations*,
PoliMI SpringerBriefs, https://doi.org/10.1007/978-3-030-34572-3_9

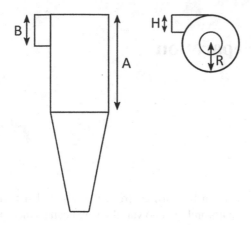

Fig. 9.1 Cyclone separator

Table 9.1 Particle-size distribution

D_p	(%)
< 3	2
from 3 to 5	8.5
from 5 to 10	10
from 10 to 15	17
from 15 to 25	25
from 25 to 30	18
from 30 to 35	12
> 35	7.5

Table 9.2 Geometrical and physical properties

Size	Symbol	Value	Unit
Radius	R	0.8	m
Inlet height	B	0.45	m
Cyclone height	A	1.35	m
Inlet width	H	0.15	m
Property	Symbol	Value	Unit
Gas density	ρ_g	1	kg/m^3
Particle density	ρ_s	2000	kg/m^3
Viscosity	μ	$1 \cdot 10^{-5}$	Pa $\cdot$ s

$$V_C = \frac{Q}{B \cdot H} = 3.7 \, \text{m/s} \qquad (9.1.1)$$

$$N = \frac{A}{B} = 3 \qquad (9.1.2)$$

$$L = 2 \cdot \pi \cdot (R - H) \cdot N = 12.25 \, \text{m} \qquad (9.1.3)$$

The smallest circumference has been selected in order to be conservative. The shortest possible path indeed corresponds to the shortest residence time for the particle after whom it cannot be collected anymore on the external wall.

On the other hand the radial distance is equal to H and the drift velocity is given by:

$$V_D = \frac{(\rho_s - \rho_g) \cdot D_P^2 \cdot a_C}{18 \cdot \mu} = \frac{(\rho_s - \rho_g) \cdot D_P^2 \cdot V_C^2}{18 \cdot \mu \cdot R} \qquad (9.1.4)$$

as a result from the equality between the centrifugal force and the viscous friction under steady state conditions. Even in this case the time required by the particle to approach the external wall is overestimated since the lowest possible centrifugal acceleration has been considered (i.e. $r = R$).

We can then match the two corresponding times:

$$\frac{L}{V_C} = \frac{H}{V_D} \iff \frac{2 \cdot \pi \cdot (R - H) \cdot N}{V_C} = \frac{H \cdot 18 \cdot \mu \cdot R}{(\rho_s - \rho_g) \cdot D_P^2 \cdot V_C^2} \iff \qquad (9.1.5)$$

$$\iff D_P^2 = \frac{9 \cdot \mu \cdot H \cdot R}{(\rho_s - \rho_g) \cdot \pi \cdot N \cdot V_C \cdot (R - H)} \qquad (9.1.6)$$

By substituting values into the equation we finally obtain $D_P^{min} = 1.54 \cdot 10^{-5} \, \text{m} = 15.4 \, \mu$. The separation efficiency consists of the particles fraction collected from the fluid phase, assuming that all and only the particles with $D_P > D_P^{min}$ are separated with 100% efficiency. Then, in this case, we should sum up the particles fraction whose diameter is higher than the minimum one (the range [15; 25] μ will be taken into account), obtaining an efficiency $\eta = 62.5\%$.

9.2 Abatement Chamber

An additional comparison between cyclone and abatement chamber having the same separation efficiency could be of interest.

The D_P^{min} is then set to $1.54 \cdot 10^{-5} \, \text{m}$ as well and the width of the chamber is assumed equal to the cyclone diameter in order to make easier the comparison of the equipment size.

The convective velocity is given by the same Eq. 9.1.1, while the drifting velocity in this case has a different driving force, i.e. the gravitational one. Its formula states then as:

$$V_D = \frac{(\rho_s - \rho_g) \cdot D_P^2 \cdot g}{18 \cdot \mu} \qquad (9.2.1)$$

Hence the two times can be set equal to each other as usual:

$$\frac{L}{V_C} = \frac{H}{V_D} \iff \frac{L \cdot B \cdot \cancel{H}}{Q} = \frac{\cancel{H} \cdot 18 \cdot \mu}{(\rho_s - \rho_g) \cdot D_P^2 \cdot g} \iff L = \frac{18 \cdot \mu \cdot Q}{(\rho_s - \rho_g) \cdot D_P^2 \cdot g \cdot B}$$
$$(9.2.2)$$

obtaining this way an expression of the chamber length as a function of the other parameters.

For the same operating conditions and efficiency of the cyclone separator, $L = 6$ m is obtained. The cyclone sizes were 1.6×1.45 for the cylindrical part, that is 1.6×2.5 taking into account the conical part, while the abatement chamber size is 1.6×6, i.e. more than the double.

This example clearly shows why cyclones are by far the most spread gas solid separator in chemical plants: they are compact and they take up space along the vertical direction requiring a smaller area in the plant.

In order to have the same encumbrance we should design a multi-layer abatement chamber with at least 3 layers and then size the height taking into account the maximum value allowed in order to avoid particles interactions.

9.3   Conclusions

Gas-solid separations are performed by means of several units, among which the most widespread are cyclones and abatement chambers. Their design has been introduced in two case studies and the results have been compared to each other.

Since these units have no moving parts, no control is possible once the design is performed. The only parameter which can be manipulated is the inlet stream flowrate, by acting on the stream itself or by using multiple units in series/parallel configurations.

Chapter 10
Centrifugal Sedimentation

Abstract Sedimentation is a separation technology alternative to filtration for liquid-solid separation. By taking advantage of different inertial behavior the centrifugal field allows for the solid deposition on the sedimenter wall and for a continuous clarified liquid removal. A multi-layered configuration, within the feasibility limits, enhances the separation effectiveness as shown in the design problem.

Keywords Suspension · Clarification · Multi-layer

Centrifugal sedimentation is a separation process aimed to obtain a clarified liquid stream (or two in case of heavy and light liquid phases) and a denser sludge from a slurry feed by mean of the different inertia of solid particles and liquid phase under a centrifugal force field.

If solid particles are big enough, the centrifugal field is sufficient to separate the two phases. The main issue could be the continous removal of the collected solid that is usually withdrawn peripherally and the clarified liquid that flows out from the top given the higher pressure of the chamber with respect to the external pressure.

If the particle diameter is too small a multilayer centrifugal decanter is used. The particle is collected when its trajectory hits the upper disc causing the solid to lose its kinetic energy. A simplified scheme of a centrifugal sedimenter is shown in Fig. 10.1.

10.1 Minimum Collected Particles Diameter Assessment

Given a vertical centrifugal sedimenter loaded with a slurry, the evaluation of the minimum collected particles diameter is requested. Both sedimenter and slurry properties are listed in Table 10.1.

Solution

The centrifugal sedimenter model can be simplified as shown in Fig. 10.2.

© The Author(s), under exclusive license to Springer Nature Switzerland AG 2020   89
A. Di Pretoro and F. Manenti, *Non-conventional Unit Operations*,
PoliMI SpringerBriefs, https://doi.org/10.1007/978-3-030-34572-3_10

Fig. 10.1 Centrifugal
sedimenter

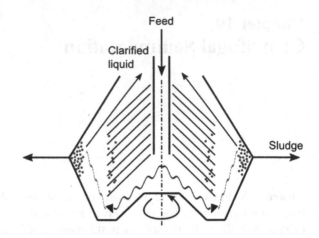

Table 10.1 Geometrical and physical properties

Physical data	Symbol	Value	Unit
Slurry specific flowrate	Q/n	0.001	m^3/s
Particle density	ρ_s	2200	kg/m^3
Liquid density	ρ_l	1000	kg/m^3
Viscosity	μ	0.001	Pa · s
Sedimenter sizes			
Discs distance	H	0.011	m
Disc angle	φ	$\frac{\pi}{4}$	rad
Disc radial coordinates	x_1	0.01	m
	x_2	0.2	m
Angular velocity	ω	60	rad/s

The particle is collected if it hits the upper disc before passing x_1. To be conservative calculations should be made for the less favorable condition that is a distance from the upper disc equal to the entire distance between two discs:

$$h = H \cdot sin\phi \qquad (10.1.1)$$

The convective velocity is obtained, as usual, by the ratio between the flowrate and the cross sectional area:

$$V_C = \frac{Q}{n \cdot B \cdot h} = \frac{Q}{n \cdot 2 \cdot \pi \cdot x \cdot sin\varphi \cdot H \cdot sin\varphi} = \frac{K_C}{x} \qquad (10.1.2)$$

Fig. 10.2 Centrifugal
sedimentation model

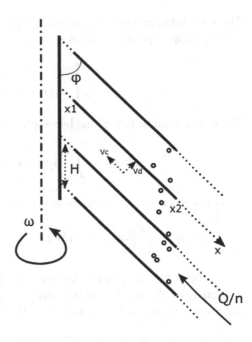

where

$$K_C = \frac{Q}{n \cdot 2 \cdot \pi \cdot II \cdot (sin\varphi)^2} \tag{10.1.3}$$

On the other hand the component of the drift velocity acting in the perpendicular direction is given by:

$$V_D = V_{stokes} \cdot cos\varphi \tag{10.1.4}$$

where

$$V_{stokes} = \frac{(\rho_s - \rho_l) \cdot D_P^2 \cdot \omega^2 \cdot x \cdot sin\varphi}{18 \cdot \mu} \tag{10.1.5}$$

that finally becomes:

$$V_D = \frac{(\rho_s - \rho_l) \cdot D_P^2 \cdot \omega^2 \cdot sin\varphi \cdot cos\varphi}{18 \cdot \mu} \cdot x = K_D \cdot x = K'_D \cdot D_P^2 \cdot x \tag{10.1.6}$$

Once we know both the velocities, the equations of motion can be written:

$$\begin{cases} \frac{dx}{dt} = -V_C \\ \frac{dy}{dt} = V_D \end{cases} \tag{10.1.7}$$

The ratio between the two components gives rise to a new differential equation. Adding an appropriate condition it results in the following Cauchy problem:

$$\begin{cases} \frac{dx}{dy} = -\frac{K_C}{K_D} \cdot \frac{1}{x^2} \\ y(x_2) - y(x_1) = H \end{cases} \tag{10.1.8}$$

The solution can be then found by simple integration

$$\int_{x_1}^{x_2} -\frac{K_D}{K_C} \cdot x^2 \cdot dx = \int_{y_1}^{y_2} dy \Longleftrightarrow -H = \frac{K_D}{3 \cdot K_C} \cdot (x_2^3 - x_1^3) \tag{10.1.9}$$

Making explicit K_D as a function of the particle diameter the solution is straightforward:

$$H = \frac{K_D}{3 \cdot K_C} \cdot (x_2^3 - x_1^3) \Longleftrightarrow D_P^2 = \frac{3 \cdot K_C \cdot H}{K_D' \cdot (x_2^3 - x_1^3)} \tag{10.1.10}$$

In order to have a deeper understanding of the relationship between the minimum collected particle diameter and the other variables of the system, K_C and K_D can be rewritten as shown in the previous Eqs. 10.1.2 and 10.1.6:

$$D_P = \sqrt{\frac{3 \cdot Q \cdot H}{n \cdot 2 \cdot \pi \cdot H \cdot (sin\varphi)^2} \cdot \frac{18 \cdot \mu}{(\rho_s - \rho_l) \cdot \omega^2 \cdot sin\varphi \cdot cos\varphi \cdot (x_2^3 - x_1^3)}} \Longleftrightarrow \tag{10.1.11}$$

$$\Longleftrightarrow D_P = \frac{3}{\omega \cdot sin\varphi} \cdot \sqrt{\frac{6 \cdot Q \cdot \mu}{n \cdot \pi \cdot (\rho_s - \rho_l) \cdot (x_2^3 - x_1^3) \cdot sin(2\varphi)}} \tag{10.1.12}$$

It is evident that the minimum diameter does not depend on the distance between the discs H as well as in the abatement chamber it doesn't depend on the chamber height. Moreover its value can be easily modulated by changing the angular speed with an inversely proportional trend. On the other hand the same observations and remarks related to the multilayered abatement chambers apply to the number of discs. Finally the numerical solution of the problem is given by $D_P = 3.15 \cdot 10^{-5}$ m.

10.2   Conclusions

Even if designed with the same purpose, centrifugal sedimentation differs from centrifugal filtration for several reasons. For instance, it can be used for denser slurries that may quickly clog an eventual filter. Moreover, it is a continuous operation and both solid and liquid phases are continuously removed, while filtration requires a maintenance dead time to clean the filter and charge the liquid hold up that has to be processed.

From a modeling point of view, centrifugal sedimentation results more similar to the abatement chambers as shown in the exercise. Furthermore, the obtained numerical results allows for the analysis of the separation performance in case of multi-layered configuration. The same restriction of the abatement chamber previously discussed applies in presence of a multi-layer centrifugal sedimenter.

Chapter 11
Centrifugal Decantation

Abstract Two liquid phases can be separated by decantation taking advantage of their different densities. When residence time should be lowered down centrifugal force instead of gravitational one can be employed. The centrifugal decanter design is based on a pressure balance and the outlet can be properly located in order to allow for a continuous operation.

Keywords Liquid-liquid · Density difference · Interface · Continuous

As well as solid-liquid separation, two liquid phases with different density can be separated taking advantage of their different inertial behaviour under a force field. This is the case of centrifugal decantation where a multiphase liquid mixture is continuously fed at the center of a fixed basket. A rotating conical disk is used to apply the desired velocity to the mixture. If the two phases are not completely miscible when they undergo a centrifugal force they separate forming two concentrical rings of which the heaviest phase is the outer one and the lighest phase the inner one.

By knowing the position of the interface between the two liquid phases, the outlet connections can be properly located in order to continuously withdraw the purified liquid.

Even in this case the separation can be enhanced by using a multi-layer centrifugal decanter.

11.1 Design of a Centrifugal Liquid-Liquid Decantation

A centrifugal decanter is continuously loaded with an oil-water mixture according to the data listed in Table 11.1. The design of the light liquid outlet is requested.

© The Author(s), under exclusive license to Springer Nature Switzerland AG 2020   95
A. Di Pretoro and F. Manenti, *Non-conventional Unit Operations*,
PoliMI SpringerBriefs, https://doi.org/10.1007/978-3-030-34572-3_11

Table 11.1 Geometrical and physical properties

Physical data	Symbol	Value	Unit
Liquid mixture flowrate	Q_{tot}^w	0.5	kg/s
Oil density	ρ_1	890	kg/m^3
Water density	ρ_2	1000	kg/m^3
Oil mass fraction	x_1	0.33	kg/kg
Water mass fraction	x_2	0.67	kg/kg
Decanter sizes			
Outlet diameter	$d_1 = d_2$	0.01	m
Total radius	R_t	0.5	m
Angular velocity	ω	10	rad/s

Fig. 11.1 Centrifugal decantation model

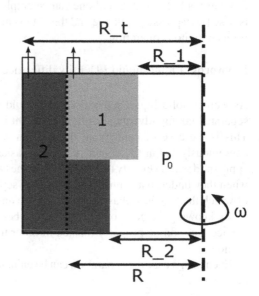

Solution I

A simple liquid-liquid centrifugal separator can be represented as shown in Fig. 11.1.

The different liquid distribution is due to the different densities, i.e. the different pressure at the interface. The pressure difference with respect to the radial coordinate can be expressed by mean of the differential equation:

$$\frac{dP}{dr} = \rho \cdot a = \rho \cdot \omega^2 \cdot r \qquad (11.1.1)$$

whose integral gives:

$$P(R_a) - P(R_b) = \frac{1}{2} \cdot \rho \cdot \omega^2 \cdot R_a^2 - \frac{1}{2} \cdot \rho \cdot \omega^2 \cdot R_b^2 \qquad (11.1.2)$$

that is nothing but the Bernoulli equation.
Pressure at the interface results then to be:

$$P_I(R) - P_0 = \frac{1}{2} \cdot \rho_1 \cdot \omega^2 \cdot (R^2 - R_1^2) \qquad (11.1.3)$$

if referred to R_1, or:

$$P_I(R) - P_T(R_T) = \frac{1}{2} \cdot \rho_2 \cdot \omega^2 \cdot (R^2 - R_T^2) \qquad (11.1.4)$$

if referred to R_T. These two values have to be the same, that means the pressure on the basket wall can be calculated by imposing their equality as:

$$P_T(R_T) = P_0 + \frac{1}{2} \cdot \rho_1 \cdot \omega^2 \cdot (R^2 - R_1^2) + \frac{1}{2} \cdot \rho_2 \cdot \omega^2 \cdot (R_T^2 - R^2) \qquad (11.1.5)$$

On the other hand, the kinetic energy balances between the outlet sections and R_T or R_1 respectively for the heavy and for the light phase are given by:

$$\begin{cases} \frac{1}{2} \cdot \rho_1 \cdot v_1^2 + P_0 = \frac{1}{2} \cdot \rho_1 \cdot \omega^2 \cdot (R^2 - R_1^2) + P_0 \\ \frac{1}{2} \cdot \rho_2 \cdot v_2^2 + P_0 = P_T \end{cases} \qquad (11.1.6)$$

Before solving the system of equations, outlet velocities need to be evaluated. The outlet cross section is:

$$S = \frac{\pi \cdot d^2}{4} = 7.85 \cdot 10^{-5}\,\text{m}^2 \qquad (11.1.7)$$

The partial volumetric flowrates can be obtained as:

$$Q_1 = \frac{x_1 \cdot Q_{tot}^w}{\rho_1} = 1.85 \cdot 10^{-4}\,\text{m}^3/\text{s} \qquad (11.1.8)$$

$$Q_2 = \frac{x_2 \cdot Q_{tot}^w}{\rho_2} = 3.35 \cdot 10^{-4}\,\text{m}^3/\text{s} \qquad (11.1.9)$$

Outlet velocities result then to be:

$$v_1 = \frac{Q_1}{S} = 2.36\,\text{m/s} \qquad (11.1.10)$$

$$v_2 = \frac{Q_2}{S} = 4.26 \, \text{m/s} \tag{11.1.11}$$

By substituting them into the system (11.1.6) and by expressing P_T as shown in (11.1.5) we finally have:

$$\begin{cases} \frac{Q_1^2}{S^2 \cdot \omega^2} = R^2 - R_1^2 \\ \frac{1}{2} \cdot \rho_2 \cdot \frac{Q_2^2}{S^2} + \cancel{P_a} = \cancel{P_a} + \frac{1}{2} \cdot \rho_1 \cdot \omega^2 \cdot (R^2 - R_1^2) + \frac{1}{2} \cdot \rho_2 \cdot \omega^2 \cdot (R_T^2 - R^2) \end{cases} \Longleftrightarrow$$

$$\begin{cases} \frac{Q_1^2}{S^2 \cdot \omega^2} = R^2 - R_1^2 \\ \frac{Q_2^2}{S^2 \cdot \omega^2} = \frac{\rho_1}{\rho_2} \cdot (R^2 - R_1^2) + (R_T^2 - R^2) \end{cases} \Longleftrightarrow$$

$$\begin{cases} \frac{Q_1^2}{S^2 \cdot \omega^2} = R^2 - R_1^2 \\ \frac{Q_2^2}{S^2 \cdot \omega^2} = \frac{\rho_1}{\rho_2} \cdot \left(\frac{Q_1^2}{S^2 \cdot \omega^2} \right) + (R_T^2 - R^2) \end{cases} \tag{11.1.12}$$

The second equation can then be solved for R and the first equation for R_1 (optional) finding this way $R = 0.343$ m and $R_1 = 0.249$ m.

For the sake of completeness the position of R_2 can be assessed by mean of the kinetic energy balance between R_2 and R:

$$\frac{1}{2} \cdot \rho_1 \cdot v_1^2 = \frac{1}{2} \cdot \rho_2 \cdot \omega^2 \cdot (R^2 - R_2^2) \Longleftrightarrow R_2^2 = R^2 - \frac{\rho_1}{\rho_2} \cdot \frac{Q_1^2}{S^2 \cdot \omega^2} \tag{11.1.13}$$

that gives $R_2 = 0.2609$ m.

In conclusion it's worth noticing that the operating parameters on which the interface position depends are the inlet flowrates and the size of the outlet channels.

Solution II

The first part of the exercise concerned the design of the light phase outlet position for a given channel diameter. However, it could be interesting to evaluate how the channel diameter affects the interface position. From a physical point of view the separation is possible as long as the interface radial coordinate is lower than the basket radius, that is as long as there's a pressure difference between the point R and R_T as suggested by the already proved equation:

$$P_I(R) - P_T(R_T) = \frac{1}{2} \cdot \rho_2 \cdot \omega^2 \cdot (R^2 - R_T^2) \tag{11.1.14}$$

In order to find the limit outlet diameter for the light phase the extreme condition of pressure equality can be set:

$$P_I(R) = P_T(R_T) \tag{11.1.15}$$

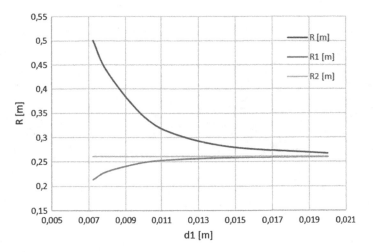

Fig. 11.2 Interface position versus outlet diameter

corresponding to:

$$\frac{1}{2} \cdot \rho_1 \cdot v_1^2 = \frac{1}{2} \cdot \rho_2 \cdot v_2^2 \iff \frac{Q_1^2}{S_1^2 \cdot \omega^2} = \frac{\rho_2}{\rho_1} \cdot \frac{Q_2^2}{S_2^2 \cdot \omega^2} \qquad (11.1.16)$$

Keeping unchanged the diameter value for the heavy phase outlet we have:

$$S_1^2 = \frac{\rho_1}{\rho_2} \cdot \frac{Q_1^2}{Q_2^2} \cdot S_2^2 = 4.1 \cdot 10^{-5}\,\mathrm{m}^2 \qquad (11.1.17)$$

that is $d_1 = 7.22 \cdot 10^{-3}$ m.

The complete interface position trend with respect to the outlet channel diameter is plotted in Fig. 11.2.

11.2 Conclusions

Although seldom discussed, centrifugal decantation is by far the most widespread operation used to mechanically separate two liquid phases; it also allows to process higher flowrates with respect to the gravitational one. The operation feasibility is based on the different inertia of the two phases, that is their different density.

As shown by the example, the unit modeling is simple and requires the solution of pressure balances only. However, the decanter size, the outlet position and the rotational speed play a critical role on a good separation without impurities; thus high accuracy during the design phase is needed even for such a simple operation.

Chapter 12
Membrane Separation

Abstract Membrane separation is usually the most refined downstream process and it is employed only if necessary due to its expensiveness. Membranes can be considered permeable or semi-permeable phase that selectively absorb and diffuse certain components. Even if the unit design procedure is case specific, it is nevertheless based on a compromise between the most severe operating conditions and the maximum allowed mechanical load as shown in the practical problem.

Keywords Permeation · Selectivity · Retention · Tubular

Solid-fluid or fluid-fluid separation can be also performed with the use of membranes due to their remarkable efficiency. A membrane is a permeable or semi-permeable phase, often a thin polymeric solid, which restricts the motion of certain species. It is conceived as a liquid phase that selectively absorbs and diffuses some components giving one product depleted in certain components and a second product concentrated in these components.

The involved phenomena depend on the membrane pore sizes. By decreasing the pore diameter the predominant separation phenomenon goes from solids mechanical interception to pressure and concentration gradients across the media. Given the general scheme in Fig. 12.1, the performance of a membrane is mainly defined by two simple factors:

- Permeation rate: the volumetric (mass or molar) flowrate of fluid passing through the membrane per unit area of membrane per unit time.
- Selectivity:

  - for solutes and particulates in liquids and gases the parameter related to the membrane selectivity is the retention. It is defined as the fraction of solute in the feed retained by the membrane:

$$Ret[\%] = 100 \cdot \frac{C_R - C_P}{C_R} \qquad (12.0.1)$$

A. Di Pretoro and F. Manenti, *Non-conventional Unit Operations*,
PoliMI SpringerBriefs, https://doi.org/10.1007/978-3-030-34572-3_12

Fig. 12.1 General
membrane scheme

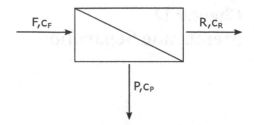

– for mixtures of miscible liquids and gases the parameter related to the membrane
selectivity is the separation factor. It is defined as the ratio of the concentration
in the permeate divided by that in the feed for two components:

$$\alpha_{A/B} = \frac{\frac{x_A}{x_B}\big|_{Permeate}}{\frac{x_A}{x_B}\big|_{Retentate}} \qquad (12.0.2)$$

In general retention doesn't depend on the component nature but only on its molecular
weight; for every membrane there exists a minimum molecular weight after which
retention is 100% (cut-off).

The pressure difference across the membrane is related both to the ΔP that should
be ensured to overcome the pressure drops due to the passage in the pores and
to the different concentration on the two sides of the membrane due to pressure,
concentration and chemical potential gradients. In both cases the flux across the
membrane can be expressed as the product between the membrane permeability and
a pressure gradient.

Membranes are sensitive units; although they are used over a mechanical support
the membrane itself should be resistent enough to avoid local damage. They are suitable for operating temperature lower than 150 °C (e.g. food industry application) and
they lose their selectivity and/or permeability due to poisoning, aging or polarization.
Polarization consists of the deposition of a superficial layer of the counter-ions of
the species for which the membrane is permeable. This phenomenon occurs during
the first hours of operation, that's why the suppliers provide as specification the permeability of the polarized membrane. In case of poisoning regeneration is difficult
or sometimes impossible causing the membrane to be a single-use unit.

Membrane filtration is a once-through operation; the employment in series of
several membranes with the same selectivity under the same operating conditions
is useless. In order to carry out multiple passes operating conditions, such as feed
pressure, should be modified to compensate the driving force decrease. They are
extremely efficient and their use requires low energy costs. On the other hand the
investment costs are higher since membrane poisoning or breaking would lead to its
substitution.

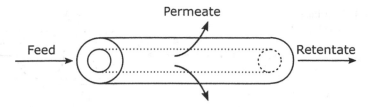

Fig. 12.2 Tubular membrane hollow fiber

Table 12.1 Membrane properties

Property	Value	Unit
N_2 permeability	$5 \cdot 10^{-4}$	kmol/(m$^2 \cdot$ h $\cdot$ atm)
O_2 permeability	$4 \cdot Perm_{N_2}$	kmol/(m$^2 \cdot$ h $\cdot$ atm)
Radius	1.27	cm
Tubes	150	1
Module length	4	m

12.1 Design of a Tubular Membrane

In case of low flowrate to process, the air separation into two streams respectively enriched in O_2 and N_2 can be performed by mean of a selective tubular membrane made up of multiple hollow fibers as shown in Fig. 12.2.

The membrane properties provided by the supply company are listed in Table 12.1.

A 46 atm pressure difference is provided at the feed inlet; on the permeate side pressure can be considered constant at 1.5 atm while pressure decrease due to the gas outflow should be taken into account on the fiber inner side. Other pressure drops can be neglected.

The number of modules required to concentrate nitrogen up to 90 % mol/mol purity is requested. Moreover, in case only 4 tubular modules are available the feed pressure needed to achieve the same specification should be evaluated.

Solution I

To correctly design a membrane separation process, the diffusive flow across the media should be properly defined. Given each component permeability the diffusive flow through the membrane can be expressed as:

$$J_A = Perm_A \cdot (P_{A,in} - P_{A,ext}) \qquad (12.1.1)$$

where $P_{A,in}$ and $P_{A,ext}$ are the A component partial pressure on the inner and the outer side of the membrane respectively.

Fig. 12.3 Infinitesimal
cylindrical volume

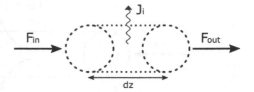

In case of a cylindrical geometry the partial mass balance for the general
i-th component can be written referring to the infinitesimal cylindrical volume
(cf. Fig. 12.3):

$$F_i^{in} - F_i^{out} = -J_i \cdot N \cdot dS \tag{12.1.2}$$

where N is the number of hollow fibers of each membrane module. Taking into
account that $dS = 2 \cdot \pi \cdot r \cdot dz$ and substituting J_i with expression (12.1.1) it results:

$$\frac{dF_i}{dz} = -Perm_i \cdot (P_{in} \cdot x_{i,in} - P_{ext} \cdot x_{i,ext}) \cdot 2 \cdot \pi \cdot r \cdot N \tag{12.1.3}$$

The model equation thus obtained is typical of any tubular shaped unit (e.g. shell and
tube heat exchanger, plug flow reactor etc.). Since the balance refers to molar flows,
they have to be made explicit in the molar fraction expressions as:

$$x_i = \frac{F_i}{F_{tot}} \tag{12.1.4}$$

Recalling that air composition can be simply assumed as $x_{O_2} = 0.21$ and $x_{N_2} = 0.79$,
the ODE system initial conditions are easily obtained.

The other variable worth to be discussed is pressure. One of the problem assumptions says that P_{ext} can be considered constant. On the other hand P_{in} decreases
because part of the gas is flowing out. The ideal gas EoS states that:

$$P_i = \frac{n_i \cdot R \cdot T}{V} \tag{12.1.5}$$

Since neither temperature nor the hollow fiber volume vary, the pressure at the general
axial coordinate z can be written by referring to the feed one as:

$$\frac{P|_z}{P_{feed}} = \frac{F_{tot}|_z}{F_{tot,feed}} \tag{12.1.6}$$

Finally, molar fractions on the external side of the membrane can be derived by
mean of a simple mass balance:

$$x_{i,ext} = \frac{F_i^0 - F_i|_z}{F_{tot,feed} - F_{tot}|_z} \tag{12.1.7}$$

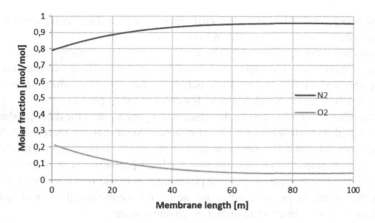

Fig. 12.4 Molar fractions profile on the internal side

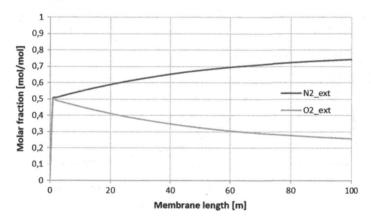

Fig. 12.5 Molar fractions profile on the external side

Since the integration domain is unknown it is convenient to integrate the ODE system over a very large interval so that the value corresponding to the desired specification can be easily detected.

The concentration profiles are plotted for the internal and external side in Figs. 12.4 and 12.5 respectively.

The total membrane length required to concentrate nitrogen up to 90 % mol/mol results then to be 24 m, i.e. 6 tubular modules are needed.

Solution II

In order to solve the second request a sensitivity analysis with respect to the feed pressure has to be performed. In the first part of the problem 6 modules were required for a 47.5 atm feed; since in this case only 4 modules are available the solution pressure is expected to be higher than this value. For the sake of completeness lower feed pressure values have been taken into account as well.

Results are shown in Fig. 12.6. The feed pressure value corresponding to a 16 m membrane length (i.e. 4 modules) is 69 atm. It is worth noticing that pressure drops considerably reduce the driving force across the membrane resulting in an infinite surface area (i.e. membrane length) by lowering the feed pressure.

On the other hand the retentate flowrate, that is a measure of the daily productivity, can be plotted as well (cf. Fig. 12.7). It results evident that, by increasing the feed pressure both capital costs decrease and productivity improvement are obtained. However, compression is expensive and for high pressure values operating costs could be a very relevant part of the total process costs. In addition it's worth remarking that, by increasing the pressure difference across the membrane, the risk of damaging the unit rises as well due to the higher mechanical stress.

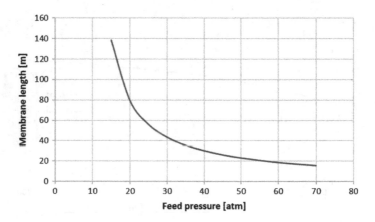

Fig. 12.6 Required membrane length versus feed pressure

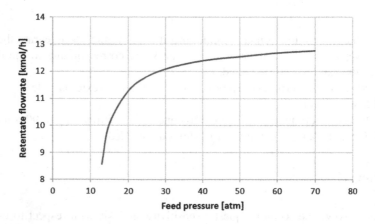

Fig. 12.7 Retentate flowrate versus feed pressure

To sum up, all those observations lead to the conclusion that there exists an optimal design solution. This operational degree of freedom can be fulfilled by mean of both an economics and safety based optimization that provides the most profitable operating conditions taking into account the unit degradation and substitution as well.

12.2 Conclusions

Beside discussing in detail the standard design procedure for a membrane separation process, the chapter provides also several remarks about operations performed with tubular units in general.

Although optimal design and operating conditions exist for the membrane tubular units as well, unlike other processes with similar characteristics the unit damage plays the main role during the decision making phase.

It is finally worth reminding that, even if a gas stream case study was considered, the domain of membrane applications is huge and could involve solid and liquid phases as well.

12.3 C++ Code

```
#define BZZ_COMPILER 3
#include <BzzMath.hpp>

void Membrane(BzzVector &y, double z, BzzVector &dy);
FILE *res;

//Operating conditions
double Fin = 20;                    // kmol/h
double xO2in = 0.21;                // mol / mol
double Pfeed = 47.5;                // atm
double Pext = 1.5;                  // atm

// Membrane properties
double N = 150.;                    // Number of fibers
double r = 0.0127;                  // m
double PermN2 = 5 * 1e-4;           // kmol/m^2/h/atm
double PermO2 = PermN2 * 4;         // kmol/m^2/h/atm

void main(void)
{
    double length = 0.;
    double delta = 1.;
    res = fopen("Res.ris", "w");
    BzzVector y0(2, Fin*xO2in, Fin*(1-xO2in));
    printf("\nINITIAL CONDITIONS:\tFO2 = %e\tFN2 = %e",y0[1],y0[2]);
    BzzVector y, dy;
    BzzOdeNonStiff o(y0, length, Membrane);
    y = y0;
    for (int i = 0; i <= 100; i++)
    {
        length = double(i);
        printf("\nLENGTH=%d\tFO2 = %e\tFN2 = %e",i,y[1],y[2]);
        fprintf(res, "\n%e\t%e\t%e",length,y[1],y[2]);
        o.SetInitialConditions(y, length);
        y = o(length + delta);
    }

    fclose(res);
}


void Membrane(BzzVector &y, double z, BzzVector &dy) {
    // FO2 = y[1] FN2=y[2]

    double FO2_ext = Fin*xO2in - y[1];
    double FN2_ext = Fin*(1-xO2in) - y[2];

    double Ftot = y[1] + y[2];
    double Ftot_ext = (FO2_ext + FN2_ext + 1e-8);

    double xO2 = y[1] / Ftot;
    double xN2 = y[2] / Ftot;

    double xO2_ext = FO2_ext / Ftot_ext;
    double xN2_ext = FN2_ext / Ftot_ext;

    double P = Ftot / Fin*Pfeed;

    dy[1] = -PermO2*(P*xO2 - Pext*xO2_ext) * 2 * 3.14 * r * N;
    dy[2] = -PermN2*(P*xN2 - Pext*xN2_ext) * 2 * 3.14 * r * N;
}
```

Chapter 13
Crystallization

Abstract Crystallization is the non-conventional operation allowing to obtain crystals with high purity from a liquor. Taking advantage of different components solubilities, more than one pure species can be obtained from the same solution. Nucleation and growth phases can be described by dedicated models. The delta-L law derivation for the crystal growth process is explained in detail and the relationship between linear growth and growth ratio is discussed by mean of a case study.

Keywords Saturation · Growth · Nucleation · Mesh

Crystallization is a process aimed to separate one or more solid phases from a solution on the basis of their solubility. It is particularly suitable to obtain pure compounds since crystals have constant composition no matter how the solution is composed.

There are two main ways for the crystals to precipitate (cf. Fig. 13.1):

- Evaporation of the solvent;
- Cooling of the solution.

Crystals formation involves both nucleation and growth processes controlled by kinetics. Precipitation occurs only if the solution is supersaturated enough to overcome the activation energy barrier needed for formation and growth of stable crystals. Therefore a second curve, the so called "supersaturation curve", exists and it doesn't depend on the compounds but on the operating conditions only.

At industrial level the nucleation is always heterogeneous since impurities are always present. However, we don't want nucleation to overwhelm growth since we need crystals that correspond to the required specification. Thus nucleation is substituted by seeding consisting in the addition of already formed crystals to the solution in order to enhance the growth phenomenon. This way the crystals size increase can be related to the residence time and the crystallization process can be easily controlled.

A. Di Pretoro and F. Manenti, *Non-conventional Unit Operations*,
PoliMI SpringerBriefs, https://doi.org/10.1007/978-3-030-34572-3_13

Fig. 13.1 Saturation and
supersaturation curves

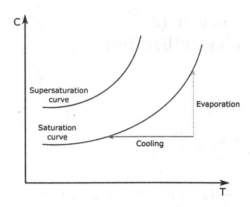

Table 13.1 Seeds
granulometric distribution

Mesh −	Mesh +	Mass fraction
	< 14	0
14	20	0.028
20	28	0.176
28	35	0.293
35	48	0.336
48	65	0.128
> 65		0.039

Table 13.2 Mesh to size
conversion table

Mesh	Size (mm)
14	1.168
20	0.833
28	0.589
35	0.417
48	0.295
65	0.208

13.1   Growth Ratio Estimation

Let's consider 1 kg of crystal seeds whose granulometric distribution is reported in
Table 13.1. Using the conversion Table 13.2 and supposing that the crystal growth
does not depend on crystal dimension (Delta-L law validity) it is namely required:

- To evaluate crystals final mass after a 0.090 mm growth;
- To calculate the growth corresponding to a final growth ratio ($\frac{W_f}{W_0}$) equal to 2.

Fig. 13.2 Concentration profile

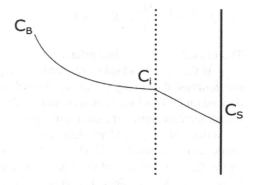

Solution I

Before going any further with the numerical solution, it is worth discussing about the theoretical background the Delta-L law is based on. The solution concentration profile can be schematized as shown in Fig. 13.2.

The reactive mass flow is defined as:

$$J_r = k_R \cdot S_C \cdot (C_i - C_S)^N \tag{13.1.1}$$

For several systems the restrictive assumption $N \simeq 1$ is very common, resulting in a straightforward experimental determination of the reaction constant. Under steady state conditions the reactive flux equals the mass transfer one according to the equation:

$$K_C \cdot (C_B - C_i) = k_R \cdot (C_i - C_S) \tag{13.1.2}$$

Thus, an overall mass transfer coefficient can be defined as:

$$\frac{1}{K} = \frac{1}{K_C} + \frac{1}{k_R} \tag{13.1.3}$$

The mass balance for each single crystal can be then formulated as follows:

$$\frac{dm_C}{dt} = K \cdot S_C \cdot (C_B - C_S) \tag{13.1.4}$$

For the crystallization process a characteristic linear dimension L is defined such as:

$$m_C = \phi \cdot L^3 \cdot \rho_s \tag{13.1.5}$$

$$S_C = \psi \cdot L^2 \tag{13.1.6}$$

where ϕ and ψ are the so called shape factors. The mass balance can be then rewritten as:

$$\rho_s \cdot \phi \cdot 3 \cdot L^2 \cdot \frac{dL}{dt} = K \cdot \psi \cdot L^2 \cdot (C_B - C_S) \Longleftrightarrow \frac{dL}{dt} = \frac{K}{3 \cdot \rho_s} \cdot \frac{\psi}{\phi} \cdot (C_B - C_S)$$

$$(13.1.7)$$

The crystal growth is then referred to the characteristic linear dimension increase only. If the right hand side terms can be supposed independent on the crystal size, the equation describing the crystal growth rate is called Delta-L law and doesn't depend on L. This implies that crystals of different size have the same characteristic length increase during the same time span.

Actually the crystal shape becomes more regular with their growth, therefore the shape factors depend on L. The reaction constant k_R depends on local conditions while K_C is a function of the Sherwood number, that is a function of L as well. However, under turbulent flow conditions, according to the relationship between Sherwood and Reynolds numbers, its value is almost constant with respect to the characteristic dimension. This is the reason why crystallization occurs in an agitated ambient in order to uniform the crystal growth despite the heterogeneous initial seeds size.

The very first step to perform is then the calculation of the corresponding size for each mesh. It is generally obtained by averaging the maximum and minimum size with the exception of the first and last meshes:

$$\bar{L}_{14}^0 = 1.168 \, \text{mm} \tag{13.1.8}$$

$$\bar{L}_i^0 = \frac{\bar{L}_i^- + \bar{L}_i^+}{2} \tag{13.1.9}$$

$$\bar{L}_{65}^0 = 0.208 \, \text{mm} \tag{13.1.10}$$

The initial seed volume is given by:

$$V_i^0 = (\bar{L}_i^0)^3 \tag{13.1.11}$$

The final crystal size is:

$$\bar{L}_i^f = \bar{L}_i^0 + \Delta L \tag{13.1.12}$$

thus the final volume results to be:

$$V_i^f = (\bar{L}_i^0 + \Delta L)^3 \tag{13.1.13}$$

Crystal density remains the same from the beginning to the end of the operation:

$$\rho_s = \frac{W_i^f}{V_i^f} = \frac{W_i^0}{V_i^0} \tag{13.1.14}$$

therefore we can write:

Table 13.3 Growth ratio vs seeds granulometric distribution

Mesh −	Mesh +	$\bar{L}_i^0$ (mm)	L_i^f (mm)	V_i^f / V_i^0	$\omega_i^0 \cdot V_i^f / V_i^0$
	<14	1.168	1.258	1.249	0
14	20	1.001	1.091	1.295	0.036
20	28	0.711	0.801	1.430	0.252
28	35	0.503	0.593	1.639	0.480
35	48	0.356	0.446	1.966	0.661
48	65	0.252	0.342	2.504	0.320
> 65		0.208	0.298	2.941	0.115
				Growth ratio	1.864

$$W^f = \sum_i W_i^f = \sum_i \frac{W_i^0 \cdot V_i^f}{V_i^0} = W^0 \cdot \sum_i \omega_i^0 \cdot \frac{(\bar{L}_i^0 + \Delta L)^3}{(\bar{L}_i^0)^3} \qquad (13.1.15)$$

where ω_i^0 is the ith mesh initial mass fraction. The growth ratio can be then straightforward evaluated according to the definition:

$$G_R = \frac{W^f}{W^0} = \sum_i \omega_i^0 \cdot \frac{(\bar{L}_i^0 + \Delta L)^3}{(\bar{L}_i^0)^3} \qquad (13.1.16)$$

The results thus obtained are listed in Table 13.3. The growth ratio results to be 1.864 and we can notice that bigger seeds affect this value more than smaller ones due to its cubic trend.

Solution II

The solution of the second part actually requires the same procedure but performed backwards. It means that, given the growth ratio, the total growth should be calculated. This evaluation cannot be easily performed by direct calculation, nonetheless the Microsoft Excel "Goal seek" function can be used as usual. This way it can be found that a growth ratio equal to 2 is obtained for a linear growth equal to 0.101; the corresponding plot is shown in Fig. 13.3.

13.2   Conclusions

This chapter provides an overview about the crystallization process and explains the derivation of the Delta L law. Although few assumptions allow for a quick solution of the problem, crystallization involves several complex mechanisms which require a deeper knowledge to be understood.

Fig. 13.3 Growth ratio
versus linear growth

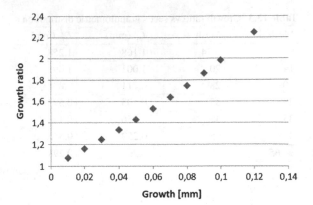

From an industrial point of view, heterogeneous seeding is always preferred to nucleation both to speed up the process and to enhance the growth phenomenon; it also has a better control on the final crystal size. In this sense the "crystal growth" parameter plays a main role in the productivity estimation.

Recommended Bibliography

1. Buzzi-Ferraris G, Manenti F (2015) Differential and differential-algebraic systems for the chemical engineer: solving numerical problems, 1st edn. Wiley-VCH, Hardcover
2. Buzzi-Ferraris G, Manenti F (2014) Nonlinear systems and optimization for the chemical engineer: solving numerical problems, 1st edn. Wiley-VCH, Hardcover
3. Flynn AM, Akashige T, Theodore L (2019) Kern's process heat transfer, 2nd edn. Wiley-Scrivener
4. Green DW, Perry RH, Southard MZ (2019) Perry's chemical engineers' handbook, 9th edn. McGraw-Hill Education, New York
5. Hundy GF (2016) Refrigeration, air conditioning and heat pumps, 5th edn. Butterworth-Heinemann
6. Haseley P, Oetjen G-W (2018) Freeze-drying, Revised edn. VCH Pub, Weinheim
7. McCabe W, Smith J, Harriott emeritus, P (2004) Unit operations of chemical engineering, 7th edn. McGraw-Hill Education, Boston
8. Myerson AS, Erdemir D, Lee AY (2019) Handbook of industrial crystallization, 3rd edn. Cambridge University Press
9. Scott K, Hughes R (2012) Industrial membrane separation technology. Springer Science & Business Media, Berlin
10. Seader JD, Henley EJ, Roper DK (2010) Separation process principles, 3rd edn. John Wiley Incorporated

Printed in the United States
By Bookmasters